D0072480

SYSTEMATIC
SYSTEMS
APPROACH

SYSTEMATIC SYSTEMS APPROACH

An Integrated Method for Solving Systems Problems

THOMAS H. ATHEY

California State Polytechnic University, Pomona

PRENTICE-HALL, INC.

Englewood Cliffs, New Jersey 07632

Library of Congress Cataloging in Publication Data

Athey, Thomas H.
    Systematic systems approach.

    Previously published as: Structured systems approach.
    Bibliography: p.
    Includes index.
    1. System analysis   I. Title.
QA402.A84   1982   003   81-15907
ISBN 0-13-880914-3   AACR2

Editorial/production supervision
    and interior design: Kathryn Gollin Marshak
Cover design: Lee Cohen
Manufacturing buyer: Gordon Osbourne

©1982, 1980, 1976, 1974 by Thomas H. Athey

Previously published under the title *Structured Systems Approach*

All rights reserved. No part of this book
may be reproduced in any form or by any means
without permission in writing from the publisher.

Printed in the United States of America

10   9   8   7   6   5   4

ISBN 0-13-880914-3

Prentice-Hall International, Inc., *London*
Prentice-Hall of Australia, Pty. Limited, *Sydney*
Prentice-Hall of Canada, Ltd., *Toronto*
Prentice-Hall of India Private Limited, *New Delhi*
Prentice-Hall of Japan, Inc., *Tokyo*
Prentice-Hall of Southeast Asia, Pte. Ltd., *Singapore*
Whitehall Books Limited, *Wellington, New Zealand*

Contents

Chapter 14   IMPLEMENTING THE SYSTEM SOLUTION   255

Chapter 15   COMMUNICATING STUDY RESULTS   273

Chapter 16   PROBLEM SENSING   291

Preface

There is a crying need today for people with very diverse backgrounds to understand systems concepts and apply them to a wide variety of situations. This can best be accomplished when people are shown the universal aspects of systems. That is, the power of systems thinking derives from viewing each situation not as if it were totally unique, but rather as a special case of more fundamental concepts which are common to all systems. In this text, you will be shown a *Systematic Systems Approach* and learn how to use this integrated method for solving complex problems. This generalized approach has been effectively applied to a wide variety of organizational problems encountered in business, government, and education, and on a more personal problem-solving level such as with groups, families, and individuals.

The primary audience for which this book is designed are those persons who are being trained through college programs to solve complex systems problems in a real-world context. These include students enrolled in undergraduate and graduate programs with schools of business that have systems or information systems majors and/or specific courses in systems analysis, systems approach, or systems concepts.

Additionally, individuals enrolled in systems thinking courses in the fields of systems science, educational planning and technology, psychology, political science, policy analysis, public administration, urban planning, military systems analysis, etc., could benefit from an understanding of the systematic systems approach. A third area would comprise those individuals presently working in organizations on complex systems problems, who need to learn a more systematic and effective problem-solving approach.

The teaching method of the text is to present an overview of the total approach and then provide a gradually more detailed explanation of the concepts, leading toward the complete systematic systems approach. This overall systems approach builds on important concepts from operations research, decision theory, economics, marketing, behavioral science, and systems theory, and results in a highly integrated framework for showing how to make systems more effective. To readers unfamiliar with these fields, some of the terms used in the text may seem esoteric—such as feasible alternatives, systems utility function, systems simulation, etc. Each term is defined first in common everyday language before it is used (e.g., feasible alternatives means workable solutions). But the terminology is kept so that the reader can relate to the concepts presented in the various fields on which this text draws.

To help illustrate the concepts of the text and to show different nuances of these ideas, there are a total of forty-five class exercises. These end-of-chapter exercises enable the student to apply the systematic systems approach concepts to problems in a personal, business, and computer information systems context. Additionally, in Chapter 17 an overall summary of the total approach is given and all steps are applied to a complete case example. To give the student additional practice in applying the methodology, a five-part hospital billing case is included. A glossary at the end contains all the terms that have been defined throughout the text.

In my doctoral dissertation, I demonstrated that this systematic systems approach is highly consistent with the characteristics needed for solving complex, real-world systems problems. Additionally, a pre- and post-test comparison was made of the thinking process of the students when they entered the course and at the conclusion. These results were then compared with control group data which had used a wide variety of other educational methods. Very importantly, individuals learning this methodology were uniquely shown to significantly increase their thinking ability, i.e., their cognitive complexity (Athey, 1976).

The systematic systems approach has been taught at both the undergraduate and graduate levels. More than 5000 people have learned this methodology at universities, community colleges, and in industrial training seminars. Based on the experience of students applying and testing these concepts on a wide variety of actual problems, and later applying these concepts in their work situations as systems analysts, managers, etc., many changes and refinements to the methodology have been added over the last four years. Further considerations that fellow instructors have suggested have also been included.

To the many people who have participated in the development of the systematic systems approach as users, teachers, or students a special thanks is due for the insight that has been gained from their experience. Appropriate reference is made throughout the text. Lastly, I would like to thank Donna Botash for typing the various editions of the text and always coming through under tight time deadlines and Karen Harvey for producing a very useful index.

Thomas H. Athey
San Juan Capistrano, California

SYSTEMATIC SYSTEMS APPROACH

Chapter One

Introduction to the Systematic Systems Approach

This book is concerned with developing a systematic method for solving systems problems. To understand what this concept really means, we need to have a precise definition of related terms like problems, solving, method, systems, and systematic. Further, we need to see how the overall process relates to other problem-solving approaches. In this chapter, we will develop a basic set of terms and present a brief overview of the *systematic systems approach.* In the following chapters we will consider in more detail each of the major points covered in this chapter.

WHAT IS PROBLEM-SOLVING?

The term problem-solving has a wide range of interpretations. Webster's dictionary defines a problem as something to be solved. Thus problem-solving would simply be deriving solutions to problems. But what are problems? We need to get some consistent definitions of problems, problem-solving, etc., before we can talk about the different types of problems and approaches.

1

Problems are situations where there is a deviation between what is expected and what actually is.

In the sense of the preceding definition for example, crossword puzzles and riddles could be considered problems. In the case of a jigsaw puzzle, assume all those little pieces of the puzzle are in a clump. What is expected or desired is to have all the pieces fit together to form the picture of a sailboat as shown on the top of the puzzle box. Since there is a deviation between what is expected and what is, there is a problem. In a similar sense, when a person has a cold, she is actually feeling sick. What is desired is for the person to feel well. The difference between feeling sick and well presents a problem. What to do about these situations is called problem-solving.

Problem-solving is the development of a solution to attain what is expected or desired from what actually is.

In the case of the jigsaw puzzle problem, problem-solving would be the procedure for putting the puzzle together correctly. The doctor's prescription for the patient is to take certain pills and rest. This will get the person well. Thus the doctor is engaging in problem-solving.

But how does one go about determining the solution to problems? What is a problem-solving method?

Problem-solving method is a specific way for proceeding in the determination of the solution(s) to a certain type of problem.

For example, we have been taught in algebra class the problem-solving method for solving a quadratic equation. In financial classes we have learned how to determine the net present value of various projects. You have learned how to bake a cake by following a certain recipe. In computer courses you were shown how to make the computer solve a particular problem by using a step-by-step approach which followed the logic of a flowchart. While these problem-solving methods are clearly very useful, each method only applies to a highly limited type of problem.

What is needed is a much more generalized approach which will guide one to the solution of a wide variety of problems. These kinds of methods are called general problem-solving methods.

General problem-solving method is a generalized way for proceeding in the determination of the solution(s) to a problem, which is applicable to more than one class of problems.

STAGES OF GENERAL PROBLEM-SOLVING

There are literally thousands of books written on particular approaches to solving very specific types of problems—for example, books on how to factor a quadratic equation, how to determine the cost of a loan, ways to lose weight, etc. What is of interest in this text is the much more generalized approach to problem-solving.

Helmholtz in 1896 put forward the idea that there are three main stages of problem-solving: *preparation, incubation,* and *illumination* (see Table 1-1). Preparation was defined in terms of seeing what the problem was. Incubation was the preparing of conditions for the development of eventual solutions. Illumination was the actual bringing forth of a solution.

Around 1938, Dewey suggested that problem-solving could be extended beyond Helmholtz's stages to include concepts such as *disturbed equilibrium, hypothesis formulation, experimental testing,* and *settled outcome.*

Disturbed equilibrium concerned the things that have caused the situation to change from the previous status quo. Hypothesis formulation is the proposal of a tentative cause for the problem. Experimental testing is the development of a procedure for verifying whether the hypothesis is true or not. Settled outcome is the resultant action taken in removing the cause of the problem and thus restoring equilibrium.

Osborne, in 1957, offered further ideas on problem-solving in his book *Applied Imagination.* He felt the stages of Dewey and others had to be further divided into the classifications of *orientation, preparation, analysis, hypothesis, incubation, synthesis,* and *verification.* Ori-

Table 1-1 Stages of general problem-solving [From Sackman (1974)].

Helmholtz (1896)	Preparation
	Incubation
	Illumination
Dewey (1938)	Disturbed equilibrium
	Hypothesis formulation
	Experimental testing
	Settled outcome
Osborne (1957)	Orientation
	Preparation
	Analysis
	Hypothesis
	Incubation
	Synthesis
	Verification

entation was required to point up the problem. Preparation was needed to gather the pertinent data. Analysis broke the relevant material
down into its fundamental aspects. The hypothesis stage generated ideal
alternative solutions. Incubation was the realization that one needs to
step back at this point, and by so doing permit an illumination of the
overall situation. Synthesis was putting all the pieces together. And the
last stage was verification, in which the resultant ideas were evaluated.

This work in defining the stages of general problem-solving was
helpful, but it had several major drawbacks. First, it wasn't very specific
so it didn't turn out to be much of a guide to problem-solving. Secondly, problem-solving needs to be applied to an actual system rather than
to theoretical problems viewed in isolation. Lastly, there must be a
specific recognition that multiple solutions are required to be considered, not just the "correct" solution.

SYSTEMS METHODS

Around the 1960s researchers such as Hitch, Quade, and Churchman developed ideas on what subsequently became known as *systems
analysis.* This incipient evolving methodology adopted the view that
problems needed to be seen in relationship to the underlying systems
they were a part of. That is, the problem of sickness needs to be related
to people systems, the problem of losing sales needs to be related to a
business system, etc. Thus there was a shift in emphasis from individual
problems to systems problems.

> **Systems problems** are deviations in systems performance between
> what is expected and what actually is.

The major difference between our initial definition of problems
and systems problems is that the deviation of what is expected from
what actually is, is measured in terms of systems performance. In the
remainder of this book we will get into the details of what systems are
and what performance means. At this stage, we can think of performance as measuring how well the objectives for the system are being
achieved.

The fundamental steps in systems analysis, as a general problem-solving method for handling systems problems, were formulated
thus: *defining objectives, designing solutions, evaluating solutions,* and
recycling until satisfied (see Table 1-2). Defining objectives emphasized
the determination of what was really desired with the system under
analysis. Designing solutions involved the aspects of developing alternatives which would meet the stated objectives of step 1. The various

Table 1-2 Stages of systems analysis as a general problem-solving method.

Systems analysis (1960s)	Define objectives
	Design solutions to meet objectives
	Evaluate cost/effectiveness of solutions
	Question objectives and assumptions
	Consider new solutions
	Redefine objectives
	Recycle until satisfactory result is achieved

solutions then needed to be evaluated in terms of how effective they were at what cost. Recycling of the steps was then done (questioning of objectives, generating new solutions, etc.) until satisfaction was reached.

While systems analysis did lead to some improvement in the general problem-solving approaches to systems problems, much was still lacking in the way of not being definitive enough nor was it complete in covering all steps in problem-solving. For example, nothing was said about the actual selection process for determining which solution is best. Further, no mention was made of the need to sell the solution, getting the solution implemented, and establishing standards of performance to insure the solution is performing "as advertised." Lastly, nothing was said about what steps to take when the new solution turns out to be in trouble.

SYSTEMS APPROACH

To answer these specific shortcomings in present problem-solving approaches, the systems approach is offered here as the best of the methods available today.

As depicted in Table 1-3, the systems approach addresses the total problem-solving cycle by including the following steps as part of this

Table 1-3 Stages of general systems problem-solving.

Systems Approach (1970s)	Formulate the problem
	Gather and evaluate information
	Develop potential solutions
	Evaluate workable solutions
	Decide on the best solution
	Communicate the system solution
	Implement the solution
	Establish performance standards

method: *formulate the problem, gather and evaluate information, develop potential solutions, evaluate workable solutions, decide on the best solution, communicate the system solution, implement the solution*, and *establish performance standards.*

Further, the systems approach includes *both* what is commonly thought of as systems analysis (formulate the problem, gather and evaluate information, develop potential solutions, and evaluate workable solutions) and decision-making (decide on the best solution, communicate the system solution, implement the solution, and establish performance standards).

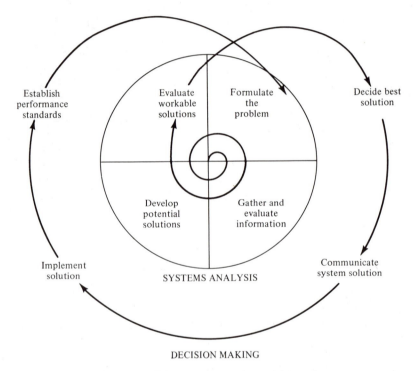

Figure 1-1   Total systems problem-solving cycle.

SYSTEMATIC SYSTEMS APPROACH

In the previous section, a complete list of the required stages of problem-solving was given. In the remainder of the text, specific chapters will be devoted to each stage—defining what the stage means and showing how to go about carrying out that phase, e.g., Chapter 4, Problem Formulation; Chapter 5, Developing System Solutions, etc.

But so far no overall scheme of attack has been given. We have not been given a guide to problem solution, only the steps that must be taken.

The guiding structure we propose is called the *system study* and is defined as follows:

Systems study is a report on what the system problem is and what should be done about it.

This type of study differs from what you are more likely to be familiar with, the research paper. Research papers, as usually assigned in college, are reportings of some situation such as, What is the situation in the Middle East?, What were the events surrounding Watergate?, What is the financial policy of the Midwestern Bank?, etc. Note however in the above definition of a system study, the part about what should be done about the situation.

Thus a study of, What is the situation in the Middle East? is a research paper. If the question is rephrased to include what to do about it, i.e., What is the best way to handle the Middle East situation looking at it from the Israeli position?, then we have a system study.

The system study in turn is made up of a *preliminary report,* a *feasibility report,* an *evaluation report,* and a *final report.* The overall purpose of each of these reports is:

Preliminary report—to determine what the problem is.

Feasibility report—to see if there are any workable solutions.

Evaluation report—to establish the relative cost/effectiveness of each solution.

Final Report—to show which solution is best and establish system goals for implementation.

The relationship between these reports and the steps in the total problem solving cycle shown in Fig. 1-1 is found in Table 1-4.

Through the use of these reports, the analyst is guided toward the best solution to a particular systems problem. As can be seen by the topics covered in the preliminary report, feasibility report, evaluation report, and the final report, an iterative approach is taken which gradually encompasses the total problem-solving cycle as a systematic way of covering more and more details, while always retaining an overview of the system.

Each of these points will be covered in much more depth in Chapter 10, What Information to Gather. At this point, it is important to note what constitutes a system study and how one goes about covering the various aspects of the problem and its solution or solutions.

Table 1-4 System study reports.

Preliminary Report
    Problem formulation
    Gather information about system

Feasibility Report
    Reformulate the problem
    Gather additional information
    Develop workable solution(s)

Evaluation Report
    Reformulate the problem
    Gather and evaluate information
    Refine workable solutions
    Evaluate advantages and disadvantages of solutions

Final Report
    Define the problem
    Gather and evaluate information
    Describe workable solutions
    Evaluate benefits and costs of solutions
    Recommend best solution
    Present solution
    Plan implementation strategy
    Establish performance standards

WHAT THE SYSTEMATIC SYSTEMS APPROACH IS NOT!

In this introductory chapter, we have tried to provide some initial insight into what the systematic systems approach is. To further aid in your understanding of this systems approach, we now need to differentiate it from other problem-solving approaches you may be familiar with. We will do this by discussing what this approach is definitely not.

Laymen as well as many academic specialists have come to use the terms operations research, management science, decision theory, computer system analysis, and economics as interchangeable with the systems approach. While the systems approach, and more specifically the systematic systems approach, draws heavily on concepts from other fields, it nevertheless differs greatly from each of the above-mentioned disciplines in that it strives for more complete solutions to a much wider variety of complex problems.

Operations research, management science, decision theory, and in fact all *quantitative business analysis* covers only the technical aspects of generating and evaluating alternative solutions to problems that have been sensed and possibly formulated someplace else. In addition to covering only a *limited portion of the total problem-solving cycle,* quantitative analysis concentrates solely on those aspects of problems which can be *quantified* and can be explicitly handled by *mathematical manipulations.*

The discipline of *systems analysis* is more general than operations research, etc., in the sense that it covers problem formulation in addition to the generation and evaluation of alternative solutions. However, it is still much more limited than the systems approach since it *does not consider any of the phases of decision-making* to be a necessary part of problem solving. What is commonly called *computer systems analysis* is even more narrow in scope than systems analysis. This analysis is limited to those problem areas which are in some way connected with the computer or *computerized applications.* While, of course, this can be a very useful thing, it is easy to see that it differs greatly from the more general scope of the systems approach.

While many who profess to be users of the systems approach have been trained in economics, it is important to realize that economics is not synonymous with the systems approach. In fact, *economics* and its sister discipline of *finance* are more limited in scope than even quantitative analysis in the sense that these disciplines have limited themselves to the generation and evaluation of solutions to those problems which can be confined to the *single dimension of dollars.* The problems of the world which meet the economics and finance criteria are thus a very small subset of those in which the systems approach is applicable.

In summary, the systematic structured systems approach as presented in this book is: (1) concerned with the determination of how to solve a very *wide variety of complex, real-world problems* and thus is not limited to computerized applications or to the economic or quantitative aspects of business-oriented problems; (2) concerned with the *total problem-solving cycle* and not just the modeling of a system or the generating of potential solutions to problems formulated elsewhere; and (3) is concerned with *effecting systems change.* A problem cannot be solved by systems studies which are not implemented. Therefore, great emphasis will be given to not only the effectiveness of solutions but also their marketability.

PURPOSE OF THE BOOK

People who can solve complex, ill-defined problems are in great demand today. As the complexity of our society increases, so will the complexity of the problems required to be solved.

Recent research in the fields of human information processing and cognitive psychology has shown that as the complexity of systems increases, the problem-solving ability of the analyst working with such systems must be greatly enhanced. Two ways of accomplishing this task are by providing individuals with methods for guiding their thinking and secondly by actually changing an individual's thinking process.

In this text, you will be learning an approach to problem-solving which has been developed and tested for its effectiveness in enhancing people's ability to solve complex real-world problems (Athey, 1976). More specifically, the reader will be shown how to:

1. Systematically approach and solve a wide variety of real-world problems.
2. Develop a user's orientation.
3. Understand the universality of systems concepts.
4. Understand the dynamic nature of systems problems.
5. Learn how to develop alternative system designs.
6. See problems in both a quantitative and qualitative perspective.
7. Gather information and evaluate it according to its potential accuracy and impact.
8. Present the results of a system study to management.
9. Understand both the technical and human aspects of implementing change.
10. See how the systematic systems approach relates to other problem-solving techniques.

Class Exercise 1-1

YOUR PROBLEM-SOLVING SKILLS

Think through some of the major decisions you have made recently, such as selecting which college to attend, what your major would be, where you should live, what kind of car you need to have, etc.

1. How did you go about solving these problems?
2. Did you use a systematic approach? Or was it more just by instinct?
3. How confident are you that you made the right choice?
4. What is the difference between problem-solving and having a general problem-solving method?
5. Think through what you are learning in various classes you are taking (e.g., accounting, management, finance, history, chemistry, etc.). Are you learning to solve problems or gaining background information, or learning new values? Relate this to the concepts of Chapter 1.

Chapter Two

Fundamental Systems Concepts

There is a clear need today for people with very diverse backgrounds who understand systems concepts and how to apply them to a wide variety of situations. This can best be accomplished if people are shown the universal aspects of systems. That is, the power of systems thinking comes not from viewing each situation as unique, but instead as a special case of more fundamental concepts which are common to all systems.

It is in this regard that General Systems Theory (GST) is being developed. One of the tenets of GST is that various properties are inherent and central to all functioning systems such as found in the mechanical, biological, and social worlds.

The purpose of this chapter is to define many of the basic properties of systems. More specifically, the common characteristics of systems will be discussed within the major areas of systems composition, systems relationships, and systems intent. Both as an aid to understanding and in order to emphasize the broad scope of systems thinking, examples will be drawn from organizations in business, government, and education. Additionally, on a more personal level families, groups, and individuals as social systems will be considered. A more detailed example illustrating these various concepts will be given in the systems context section.

SYSTEMS COMPOSITION

In this section we define what a system is and identify its basic elements to include system boundaries, environment, and system interfaces.

> **Systems** are any set of components which *could* be seen as working together for the overall objective of the whole.

An orchestra could be looked at as a system. The various parts of this system include the musicians, the various types of instruments, the musical arrangement, and the conductor. The overall objective of the orchestra, as a system, is to create an excellent musical expression.

Even a book can be considered as a system. The components include the concepts, words, sequences, formatting, cover, binding, drawings, etc. To be operative as a system, these components would be selected and fitted together in such a way as to attain an objective like entertaining or educating the reader.

Very importantly, note that in our definition of systems the qualifying phrase *could be* is used. This implies that irrespective of whether or not anyone is presently considering a set of components as a system, it still may in fact be a system!

> **Components** are the primary elements which comprise a system.

One popular system in use today is the motorhome recreation vehicle. Its components include the chasis, the type of engine, the interior layout, the refrigeration and heating apparatus, the disposal equipment, the capabilities and limitations of the people using the vehicle, the type and amount of gasoline and butane available, etc.

We note that the word components in the systems definition is used to depict a comprehensive set of elements to include people, machines, money, concepts, processes, feelings, beliefs, etc., and thus is not limited to objects.

To further emphasize the generality of systems and see how they are not in any way limited to physical objects, consider the beliefs of various religions as particular systems. For example, the ideas put forth in the Bible, the Koran, or the Book of Mormon may be seen as the set of components which is used to describe certain patterns or belief systems. The overall objective of these religious systems may be to provide guidance, or foster truth, wisdom, salvation, etc.

System boundaries comprise that set of components which can be directly influenced or controlled in a system design.

In our definition, we stated that systems include any set of components which can be looked at as working together for some overall objective of the whole. This definition does not differentiate the components which are controllable from those which are not. Systems which are part of the total system, but which can't be controlled will be considered as part of the environment (Schoderbek, et al.).

Environment includes all those factors which have an influence on the effectiveness of a system, but which are not controllable.

Consider the case of a large organization like Holiday Inn trying to decide where and what type of hotel it should build in a major international country like Israel. The management of Holiday Inn has control over what price they are willing to pay for land, what specific site to choose, what type and size of hotel to build, what kind of clientele they are trying to cater to, etc. These elements of the hotel system are within the systems boundaries, because they can be controlled or directly influenced.

For an understanding of environmental factors, note that the hotel business attracts mainly two types of customers, businessmen and vacationers. These people in turn are influenced by the general state of the economy. If the economy is going great, businessmen can attend international conferences and can travel to stimulate business deals. Likewise, consumers have the money and desire to take vacations when times are good. The exact opposite happens when the economy is in a severe recession.

Thus the future prosperity of the Holiday Inn in Israel will be greatly affected by how well the economies of their potential customer's countries are doing. Additionally, if a state of war erupts within Israel, the tourist trade would be very much changed.

In both of these cases, Holiday Inn is greatly affected, but in neither case does it have any control. Thus economic and war conditions are part of the environmental aspects of the system, and not within the systems boundaries; but the location, the size, and architecture of the hotel, etc., are within the system boundaries.

Interfaces are those boundaries where two systems meet, such that the output of one system is the input to the other. These boundaries can be internal or external to the system itself.

An example of an external interface would be an organization and its customers. Assume you worked for a local water district office which needed to bill its 100,000 customers each month. After analysis of the time, cost, and accuracy of the present manual method of billing, you decided that a computerized billing procedure would be more effective. You therefore contacted the IBM representative who suggested a medium-size computer for your organization. You agreed. The output of the IBM system is the computer. This output becomes an input resource to the water district company system.

An internal interface could be some type of assembly line where a product moves through various points in its development process. A college curriculum has many internal interfaces. For example, in a required four-course sequence in mathematics (e.g., algebra, statistics, calculus, and differential equations) each course is a prerequisite to the next course. Certain student skill levels upon leaving the algebra course are assumed, for example when the students enter the statistics course, and so on. For good student performance, the teachers (decision-makers) of each course need to coordinate the subsystem interfaces with respect to overall system objectives.

The developer of any large commercial or residential property today has a great many internal and external system interfaces which he must coordinate to insure a successful project. For example, take the development of a large shopping mall in the vicinity of a major city. The developer must work with city, county, state, and federal officials concerning environmental impact studies, building codes, safety standards, financial loan conditions, minority hiring statements, etc. In addition, the developer must coordinate work with independent plumbing, electrical, carpenters, air conditioning, cement, and trucking contractors. He must also work with major retail companies like Sears, Buffums, and May Company to insure a good mix of retail firms that will attract the buying public. To develop a prosperous shopping mall complex, each of these interfaces must be successfully negotiated which in turn requires much time, skill, and luck.

SYSTEMS RELATIONSHIPS

In this section, we consider how systems are made up of subsystems and how knowing who the decision-maker is for a particular system affects this viewpoint. Further the interrelationship of subsystems results in a system structure, which can be changed through various systems designs.

Hierarchy concerns the relative relationship between systems and their components in terms of supra- and subordination.

A fundamental property of systems—whether they are large and complex, or small and relatively simple—is that they are composed of components, which are systems in themselves and are called subsystems. In turn, what we have called a system can be looked at as a component of a larger, more encompassing system called a suprasystem.

This leads to one of the typical problems encountered in applying the systems approach to real-world problems: namely, how to decide what is the appropriate level at which to study a system, i.e., which is the system and which are the components.

For example, assume you were asked to develop a better reading program for the first graders of the ABC Elementary School in Midtown, U.S.A. What looks at first like a very well-defined problem would soon turn very complex when you realize that reading is just one phase of the first-grade program at this school. So perhaps you decide to consider the total first-grade program and then within this framework determine how best to handle the reading. However, this approach is suboptimal since the first grade is just a part of the total elementary school system. Therefore, one really needs to look at what are the objectives of the elementary school system and how the reading program best fits in this scheme. But the elementary school is a subsystem with respect to what the district school system is trying to do. The district school of Midtown, however, is just a part of what is being attempted in the educational system of the State.

The same reasoning could lead eventually to a need to consider the total educational system of the world first, for fear of suboptimizing when designing the original reading program. In fact, note that the educational system of the world is a subsystem relative to the overall objectives of what the world system is trying to accomplish.

Since one obviously cannot deal with all problems on a world-scale first, before tackling any "minor problems" for fear of suboptimizing, how is this situation to be handled? The rule that Churchman has suggested, which is very workable, is to determine who is the *chief decision-maker (CDM)*.

> **Chief decision-maker** (CDM) is the one for whom the study of the system is ultimately being done. He is the person or committee who has authority to change the system in accordance with the results of a system study.

The primary purpose of a system study is to increase the effectiveness of a system in accomplishing agreed-upon objectives. Recommendations from a system study which are not implemented, clearly don't positively affect the system. A most common reason why study results or recommendations aren't put into effect is that the analyst's conclusions go beyond the authority or control of the CDM. Therefore, to greatly increase the chances of a study changing the system, the analyst needs to look at the hierarchy of systems in terms of the CDM's control and then define the systems boundaries and the environment under these conditions.

We have defined the system and its boundaries in relationship to the CDM's authority. Anything outside the CDM's control is called environment. For example, if the CDM for the first-grade reading program is the principal of the elementary school then the system boundaries are all resources he has control over. Any requirements outside his control which the system must meet are real environmental restrictions. For instance, there may be a state law which requires all reading programs within the state to use reading books from the state-approved list. Unless one can get that list modified, he must design his solutions taking that requirement as a given even though it may not be what is best for that particular elementary school!

If one feels these restrictions are hurting the system performance greatly, he may then try to get the study approved at a higher level. By moving the CDM to a higher systems level, a greater number of factors becomes controllable, and that which was an environmental restriction is incorporated within the system boundaries.

> **Structure** depicts the relationships between the components of the system to include its organization and interactions.

A business enterprise can be considered as a system. Its components include the people, facilities, equipment, capital, and materials that it blends together in a certain way to produce various products and services valued by its customers. In so doing, the business hopes to be able to meet its overall system objectives of maximizing profit, growth, goodwill, etc.

Its organization chart would show the structure of one aspect of the business system, namely, that of the people in the company and their formal relationship to each other. The degree of organization among the company members, and the resulting frequency and intensity of their interactions would vary if the system organization were a traditional pyramidal structure as compared to the much more fluid, problem-oriented matrix structure.

A football team can be viewed as a system. It is composed of people who play the organizational positions of center, tackle, guard, end, fullback, quarterback, etc. The particular formations that the team uses on the offensive and defensive describe the structure of the system.

While all football teams have eleven persons on the field at one time, many variations in the relationships of the players are possible. Watching high school, college, and professional football games will show the great difference in the degree of organization among the players and how the professionals tend to work as much more highly developed team systems as opposed to a collection of individuals.

Systems design is concerned with the appropriate selection of system components and their arrangement (structure), so as to meet the overall objectives of the system.

An automobile can be considered as a system. Its components include the type of make and model, engine, transmission, body style, color, etc. By selecting different types of components and arranging them in different ways, many alternative system designs are possible.

For example, a 1977 Pinto two-door, with a 1600 cubic centimeter engine, automatic transmission, deluxe interior, blue, etc., is one design. Another could be a 1973 Corvette, two-door with a sun-roof, 3500 cubic centimeter engine, five-speed manual transmission, wide-track tires, candy-apple red, etc. A third alternative design could be a 1976 Mazda station wagon, with a 2000 cubic centimeter rotary engine, etc. Expanding on this example, it is easy to see that car manufacturers provide us with literally millions of different auto systems designs.

In a similar way, a computer can be considered a system. The components of hardware would include the central processing unit, the kinds of input devices, output devices, means of storage and the software considerations involving the languages available, application packages, etc.

One of the possible computer system designs available for a small business today would be a Burroughs minicomputer with a card reader used for input and a line printer for output. An alternative design would be to keep the same processing unit, but have a terminal access and augment the storage with a disk unit. A third hardware design would be to go with a much different concept such as the Radio Shack microcomputer. More specifically, attached would be several video terminals, a character printer, and a floppy disk storage device all set in a dispersed data processing mode.

For a much different concept of system design, consider the family as a social system. Women today have for possibly the first time, the choice of alternative family systems. That is, they don't have to accept the traditional systems design of marrying a man and becoming a wife and mother who stays at home with the children. Other alternatives are (1) staying single and pursuing a career; (2) marrying, having no children, and pursuing a career; (3) living with a man, having children, and staying at home, etc.

What determines whether a systems design is good or bad? The various system configurations need to be evaluated against the objectives and purposes of the system. In this context, those system designs which attain the objectives of the system to the greatest extent are considered best.

SYSTEMS INTENT

In this section, the meaning of systems as related to both objectives and purposes from an overall or wholeness point of view is discussed. The resulting conflicts and the need for making tradeoffs are emphasized.

Wholeness is concerned with the overall aspects of a system, not with the individual components, per se.

A distinguishing characteristic of systems thinking is the realization that (a) the whole is more than the sum of the parts and (b) what is best for the subsystems (components) is not necessarily what is best for the overall system or vice versa.

The first part above refers to what is called synergy. That is, the system is more than the totality of its components. For example, recall the traditional story of the chemist who took a human body and determined its chemical composition. He then went and priced these chemicals, and established that a human being is worth $1.50. This chemical analysis and the economic equivalency of those chemicals are both correct. But his conclusion that the human being is worth $1.50 because this is what the chemicals cost is wrong because it assumes the whole is the sum of the parts. If he had looked at the human being as a system, he would see that its value isn't determined just by the components. Rather what also must be considered is the interrelationship of these chemicals, which uniquely forms a living human being.

The second point also concerns the parts versus the whole. As an illustration of the case, assume you were an advisor to the Canadian government who in 1975 was trying to select a hockey team to play the touring team from Russia. There was lots of prestige riding on the outcome of this series, since Canada prides itself on having the best hockey in the world. How would you select the members to play on this team?

Canada selected their team by picking the all-stars from each of the eighteen teams in the National Hockey League. What implicit assumption did they make? By picking the best individual players at each position (i.e., components), they assumed they would come up with the best overall team (i.e., system). Team Canada got trounced by the Russians, who most likely couldn't match the talent of the Canadian team man for man, but they had the advantage of having worked and played together for many years (i.e., interrelationship among components). An alternative strategy for the Canadian government would have been to pick the top team (in 1975 it was the Philadelphia Flyers) and let them play as a team, since they had proved as an overall system that they were very good.

The overall point of this section is that an analyst needs to keep his eye on the overall system, not just the components.

Objectives are the goals or results that the decision-maker wants, or should want, to attain in regard to a particular system.

The business school or university could also be looked at as a system. The individual parts would include the courses, teachers, staff, students, buildings, etc. The overall objective of this system could be to impart knowledge to students. Or the system objective could be to prepare students for life or for a particular job. Maybe the objective of the business school system is really to give teachers a nice environment to work in. Possibly, all these objectives would be considered simultaneously. In any case, depending on which objectives are most influential, the relative importance of the students, staff, and buildings will vary greatly in the attempt to have the best overall systems design (i.e., that system configuration which best meets the overall objectives).

As another illustration, we consider the example of an airplane as a system. The components would include the engines, fuselage, wings, interior and exterior size, the type of materials used, the type of instrumentation available, etc. What should be emphasized among these components depends on what you want the airplane to do (i.e., system objectives).

How do fighter aircraft, commercial airlines, and gliders differ and why? Take the type of engines, for example. Because the fighter aircraft needs great speed and maneuverability, they have high-thrust jet engines with afterburners which enable the airplane to attain supersonic speeds. Since commercial airlines are more concerned with being able to carry many passengers, as opposed to supersonic speed, the best engines are those which provide safe and reliable power for carrying heavy loads. The glider, on the other hand, is attractive because it has no engine. Its purpose is not speed or distance, but soaring time.

Purposes of a system are determined by its relationship with the environment.

In the case of a business organization, the customers' desires establish the purpose of the particular firm. For example, the purpose of Blue Cross is to provide people (their clients) with health insurance to offset expenses incurred during illness or accidents. That is one interface; another system interface is hospitals. To various hospitals the purpose of Blue Cross could be to provide a ready supply of patients (customers), who represent a minimum risk financially.

Thus, an organizational system has as many purposes as it has interfaces. Which of these purposes it emphasizes and how it goes about satisfying them depends on the system objectives of that company. The purposes and system objectives of an organization are very different things. The system objective of Blue Cross as a business could be to maximize its gains. The way it might do that is to provide good health insurance to a very select group of individuals who are healthy and disregard the needs of people in high-risk categories or vice versa.

As another example, consider the question of why do you, as a specific individual system, wear the particular clothes you have in your wardrobe? You probably will answer, to keep warm, for modesty, and/or because when you wear sharp clothes, it makes you feel good.

Molloy, in his excellent book *Dress for Success*, gives much support for the view that people use the way you dress as an indicator of your success in life. This in turn influences the way they react to you. Assuming this is true, then the purpose of your clothes, irrespective of how you feel about it, is determined to a large extent by your environment (i.e., the people you come in contact with).

This reasoning does not imply you have to dress to satisfy others, only that you may pay a price if you don't (i.e., systems tradeoff). The views of others may be of less importance to you than your own personal independence. Recall the difference between system objectives and purpose.

However, before you dismiss this thought, consider the last time you went for a job interview, on an important date, or made a speech: Why did you get dressed up? What would the consequences have been if you hadn't?

> **System tradeoffs** involve the explicit recognition that a system can't be designed so as to satisfy the multiple, conflicting objectives and purposes of a system equally well.

Implicit in all that has been said so far is the concept of limited resources. Because men, machines, land, energy, capital, time, etc., are of limited quantity, tradeoffs are required in any system design. Conflicting demands should be resolved at the system level in terms of the *relative* desirability of various objectives and purposes.

For example, a student while attending college may desire to have a good time, learn about a wide variety of subjects, become qualified for a good job, spend lots of time skiing and swimming, have a nice car, be free of pressure and harrassments, etc. However, he soon learns that to take all the subjects he likes or is interested in will require him to go to college for six years versus four years if he took just the prescribed courses in his major. To be able to get a good job, an individual will most likely have to have a solid B average while he is in college. This in turn will require the student to spend many hours studying. Unfortunately, the time he spends studying is time he can't spend at the beach.

Or take the case of one of the most important decisions a person can make in life, selecting a career. Nearly all of us would like jobs which provide excellent pay, terrific promotion opportunities, great challenge and personal independence, unquestioned security, fine health and retirement programs, flexible work hours, rewards for personal integrity, etc. Unfortunately, there aren't any career fields which provide all of these things. Those jobs which pay well, generally don't have much security. Having both the flexibility to work when you want and terrific promotion opportunities within the same organization is very rare. Good retirement programs and personal independence are conflicting aims.

Thus, it is not a question of whether or not tradeoffs should be made—they have to be! Rather the relevant question is which are the best tradeoffs to make in regard to an individual's system, considering his present and future objectives.

SYSTEMS CONTEXT

So far we have discussed the fundamental concepts of systems composition, systems relationships, and systems intent. We now consider how to aid the analyst in seeing these concepts in regard to a particular system he or she might be studying. To help keep these elements in the proper perspective, the *systems context diagram* has been developed (see Fig. 2-1).

Systems Context Diagram

The complexity inherent in any real-world system is quite high, and thus there is a great chance the analyst will immerse himself in the details and thus not keep an overall perspective—similar to the problem of not being able to see the forest for the trees. Therefore, the analyst needs to abstract the crucial aspects of a system and see how it all fits together.

The systems context diagram depicts the system and its major parts. It also shows the interrelationships between the concepts of the

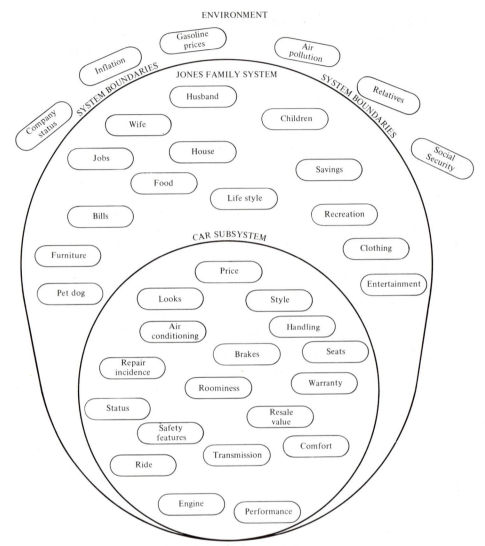

Figure 2-1 Systems context diagram.

CDM, systems boundaries, and the environment. Additionally, it shows the subsystem being studied and its components. In the next chapter, we will show how to increase the value of the diagram to the analyst.

However, this diagram is limited to a static view of the system. Any problem must be studied in the context of an overall system, not in isolation. That is, one should not ask the question, What is the best car? Per se, this question has no meaning. Rather, the question needs to be restated as, What is the best car for the Jones family? Since it now relates to the context of an overall system, the question is meaningful.

Family as a System

To see how this all works, consider the family as a system and in particular the Jones family (Fig. 2-1). The system components would include the people subsystems of a husband, a wife, and say two children. Also part of this family system would be the material subsystems of a car, a house, furniture, etc. The monetary aspects would include the present savings/asset level, gross income, credit ratings, investment potential, etc.

The objectives of this family system could be growth, happiness, security, health, independence, etc. These in turn will help determine what should be emphasized in any new system design or change of components. This should enable an analyst to conclude which are the best systems tradeoffs between money, quality of life, status, etc.

The structure would include how the family components of husband, wife, children, money, cars, job, house, etc., are interrelated and organized. That is, Is it a children-oriented family? A husband-dominated group? etc. How is the family work divided? Who does what? How was it decided? What are the means of communication and what is discussed by whom? etc.

Class Exercise 2-1

MANUFACTURING SYSTEM

Fig. 2-2, from Awad's *Introduction to Computers in Business,* is a diagram which shows a manufacturing firm as a system which is composed of subsystems. Let's use this figure to apply what has been discussed in Chapter 2.

1. What is the system? What are the subsystems?
2. What is the relationship between accounting, payroll, and withholding tax reports? What are the major interfaces?
3. What are the system boundaries? How are they determined?
4. If we consider Fig. 2-2 as a systems context diagram, what major elements are missing? Why didn't Awad include them?
5. What are the objectives of the manufacturing firm? What are the objectives of the accounting department? the payroll section? What is or should be the relationship between these various objectives?

Class Exercise 2-2

CLASSROOM AS A SYSTEM

Look at the specific classroom situation in which you are being exposed to the systematic systems approach. Use the concepts and definitions learned so far to answer the following questions.

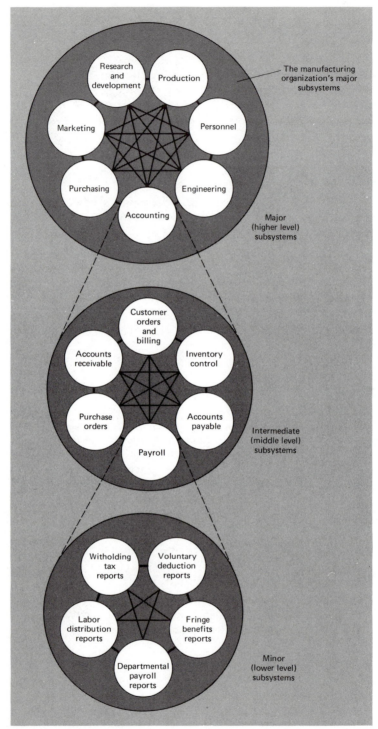

Figure 2-2 [from Awad, *Introduction to Computers in Business,* Prentice-Hall, 1977].

1. Describe the classroom as a system and elaborate on what the major components are (use Fig. 2-3 as a starting point).
2. What is the present systems design?
3. What are the system objectives?
4. Who is the CDM? What components does the decision-maker have control over, and which are given?
5. What specific demands are placed on this system from the immediate environment?
6. Develop several other plausible system designs.
7. What influence does the teacher have on the way this course is taught?

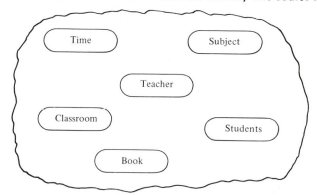

Figure 2-3 Components of an educational course.

Class Exercise 2-3
PRISON SYSTEMS

There is an extensive prison system in this country which houses criminals for a variety of reasons.

1. Why can prisons be looked at as a system? What are the components and hierarchy? Give a typical systems design.
2. What is their environment? Who is the CDM? What are the major interfaces within and outside of a prison?
3. What are the different purposes of various federal penitentiaries, state prisons, county jails, etc.? How would purposes and objectives differ depending on whether you were viewing these systems from the perspective of society, prison officials, or the prisoners themselves?
4. A prisoner in a maximum security prison is kept isolated from the outside world. Is this prisoner functioning as a static, dynamic, homeostatic, or cybernetic system? (See p. 27 for definitions.) Why is it such a cultural shock for these prisoners when they are released from prison after 20 years? Why are there frequently interim "halfway" houses?

Chapter Three

Cybernetic Systems

In the preceding chapters the basic systems concepts were defined. In this chapter we will show how the particular composition, relationships, and intent of a system can take different forms depending on what type of system is being studied. The basic types of systems will be classified as *static, dynamic, homeostatic,* or *cybernetic* depending on (1) how much influence the environment has on the system, (2) the degree of internal control the system has developed, and (3) whether the system goals are fixed or adaptive.

A case will be made that while the systematic systems approach is in no way limited to the study of cybernetic systems, in this text we will concentrate on examples of systems which are complex and social in nature, such as business, educational, and governmental organizations and, more personally, families and individuals. According to the scheme developed in this chapter, cybernetic systems are the only ones which make sense in this context. However, we will need to know the different forms systems can take, so we, as analysts, will know when a system is actually functioning in a cybernetic fashion or, equally important, how to make a system cybernetic.

CLASSIFICATION OF SYSTEMS

While the basic types of systems can be classified in many ways, one particular three dimensional display is shown in Fig. 3-1, This grouping is based on (1) the effect the environment has on the system, (2) the amount of internal control the system has, and (3) whether the system goals are fixed or adaptive.

Environmental influence concerns how much effect the environmental conditions have on the functioning of the system. If the system is independent of the environment, it is called a closed system. If the system output is very much influenced by changes in the environment, the system is called open.

Control is how much internal capacity a system has for insuring the continual attainment of system objectives. This can range from a system having no internal feedback devices to systems which have very effective feedback loops.

Adaptability of goals reflects whether the goals of the system are fixed or can be changed depending on the environmental condition and state of system learning.

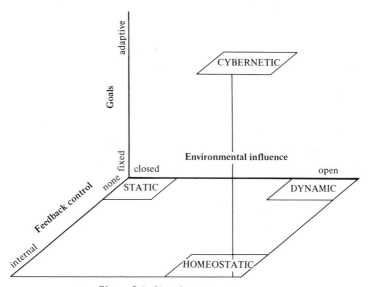

Figure 3-1 Classification of systems.

Static Systems

Static systems are those which are closed to environmental effects, have fixed goals, and have no means of internal control to insure that the system goals are met. An example of a static system is a watch (clock). The functional purpose of this system is to keep accurate time. The designer of a watch makes the various component parts blend together to best accomplish the overall systems goal. This system has no internal control devices for insuring it is keeping accurate time. The watch may gain five minutes per day or lose ten minutes per day. The watch itself has no way to readjust its systems output (time) so to be more accurate. The watch as a static system also has no awareness of environmental shifts. Assume the watch is working perfectly and is keeping accurate time. If the environment now shifts from standard time to daylight savings time, the watch as a closed system will continue to function exactly as it was designed to do. The environment in no way affects the internal workings of this system.

Dynamic Systems

Dynamic systems are those systems which have no means internally to insure that the fixed system goals are met under highly variable environmental conditions. Figure 3-2 represents this type of system schematically. 195565166 MICHAEL HAAK

An example of a dynamic system would be an inexperienced pilot flying a small private plane cross-country. This pilot desires to get from point A to point B, a distance of 100 miles, in one hour. The pilot therefore sets the throttles (amount of gasoline to engines) such that the plane will go 100 miles per hour through the air. Whether in fact the airplane will cover this distance in one hour is dependent on the environment. If there is no wind, the airplane will go from point A to point B in one hour. If there is a tail wind of 100 mph, the airplane will get to point B in ½ hour. If there is a head wind of 100 mph, the airplane will stay over point A until it runs out of gas. Therefore, the output (distance covered) of this system is strongly affected by the environment. While the system objective has remained fixed (going from point A to point B in one hour), the actual system output

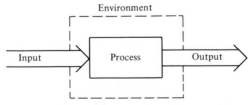

Figure 3-2 Dynamic system.

(distance covered) will be highly influenced by the environmental conditions.

If an experienced pilot were flying the aircraft, he could function as a control device to insure that the system objective was met by increasing or decreasing the speed of the aircraft.

Note however, in the case of a person planning a trip on a sailboat from Los Angeles to Hawaii that this is truly a dynamic system. Even an experienced sailor has no way of insuring he will make the trip in any specific amount of time. The speed of the sailboat is a direct function of the environment (mainly the speed and direction of the wind). If there is no wind, the sailboat isn't going anyplace!

Homeostatic Systems

Homeostatic systems are those which respond to environmental changes, but have effective internal control devices to enable the system to meet its fixed system goals (Siegel, 1969). Figure 3-3 diagrams the principle of this system.

As an example of a homeostatic system, consider the temperature control system within a house. How hot or cold the house gets is a function of two main inputs: (1) how hot or cold it is outside the house and (2) how much heat or cool air the central system is providing. These inputs are combined in the air mixing process (see Fig. 3-4). The output of this process is the resulting degree of warmth or coldness of the house. The output stream is measured and compared to an established standard (i.e., thermostat setting). If the air temperature in the house is below the standard (i.e., too cold), then a correction signal is given which turns on the heating unit. This device then adds hot air to the current air until the output mix compares exactly with the standard. At that point no more heat is needed and the heating system is turned off.

This homeostatic system functions effectively to meet the fixed

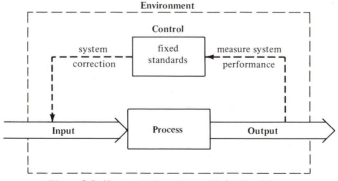

Figure 3-3 Homeostatic system with feedback loop.

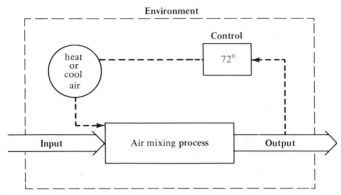

Figure 3-4 House temperature control system.

system goals. Whether or not the system goal is correct is never questioned. Somebody could accidentally have set the thermostat at 110°. The feedback system will accept that goal and continue to supply heat until the goal is attained.

To see the difference between open and closed feedback systems, consider Fig. 3-4 again but this time assume the house has perfect insulation. That is, no matter how hot or cold it is outside the house none of the external factors penetrates the walls of the house. In this case, the house temperature control system is a closed feedback system (i.e., independent of external influences). The air temperature inside the house would be a function only of (1) what activity was going on inside the house and (2) how much heat or cool air the central system was putting out. Activity within the house such as cooking, running the dryer (heat), or building a fire in the fireplace would all affect the air temperature inside the house. Important to note, however, is that all that activity is internal and within the control of the CDM (collectively, the occupants).

If the insulation in the house is less than perfect, this means that the air temperature within the house is directly influenced by conditions outside the house. The house temperature control system thus becomes an open feedback system (i.e., homeostatic).

Cybernetic Systems

Cybernetic systems are those systems which are affected by environmental shifts but have means through feedback control to continue to meet system objectives. Additionally, the system objectives are not rigidly fixed but are adaptable to changing conditions and responsive to new understanding. These systems gain from experience and thus exhibit learning.

An enlightened profit-oriented business is an example of a cybernetic system. A business firm has inputs, processes, outputs, and measures of success (see Fig. 3-5). The basic inputs into this system are the resources of material, men, capital, and information. The organization takes these inputs and through its processes outputs various products. The amount of products sold and specifically the profit derived from these sales are monitored and compared to a standard. Readjustments are made in the system in light of this comparison to better meet the overall system objective of, say, maximizing profits.

If the maximizing profit standard is fixed without regard to changing cultural and social conditions, or new learning that takes place internally to the firm, then the business is in fact acting as a homeostatic system. What makes this system cybernetic is the fact that many businesses have found that maximizing profits is not exclusively what they want or should want to do. Many companies, for example, have expanded their control standards (systems objectives) to include elements such as increasing market share and developing product leadership in addition to profits.

During the 1960s and 1970s many businesses became increasingly aware of their obligation to society (the larger system that the business is embedded in) above and beyond maximizing profits. These firms have accepted as legitimate system objectives the socially responsible goals of hiring minorities, reducing pollution, recycling materials, and so on, in addition to profit, market share, and product leadership.

Regardless of whether or not these new system objectives are the best ones, the point is made that these systems are cybernetic. They have been affected by the changing environment for businesses today and have showed adaptability by changing system objectives to reflect new understanding.

Summary

In this section, we have established one scheme for classifying systems according to environmental effects, controls, and adaptability

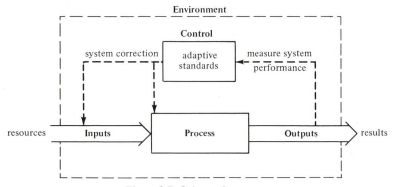

Figure 3-5 Cybernetic system.

of goals. We can apply this scheme to many types of systems, whether they are abstract, mechanical, biological, or social.

Generally speaking, abstract systems like mathematics, language, maps, etc., are usually closed systems. Mechanical systems like computers, typewriters, and cars mostly fit into the dynamic systems classification. Some sophisticated mechanical systems like the autopilot in airplanes and the guidance system in missiles are really homeostatic. Most biological systems, like plants and lower level animals, function as homeostatic systems with a fixed goal of survival.

Social systems such as an individual, group, or organization *should* function as cybernetic systems. However, we all know of examples where companies or individuals function homeostatically pursuing fixed goals, or as dynamic systems just swaying with each environment change, or even as closed systems allowing no new or disrupting information to be considered which conflicts with previously set beliefs. For a very interesting study of this latter case, see the classic book by Rokeach, *The Open and Closed Mind* (New York: Basic Books, 1960).

A BUSINESS AS A CYBERNETIC SYSTEM

An enlightened profit-oriented business is an example of a cybernetic system. A business firm has inputs, processes, outputs, and measures of system effectiveness (see Fig. 3-6). The basic inputs into this system are the resources of men, material, money, machines, and information. The organization takes these inputs and through its processes outputs various products or services. The amount of products sold and specifically the profits derived from these sales are monitored and compared to a standard. Readjustments are made in the system in light of this comparison to better meet the overall system objective of maximizing profits.

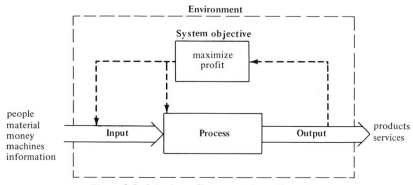

Figure 3-6 A business firm as a cybernetic system.

General Motors during the period 1974–75 is an excellent example of how a firm functions as a cybernetic system. G.M. had to develop alternative strategies in an attempt to maintain overall system effectiveness by learning how to cope with a complex and difficult environment. The existing system strategy of G.M., under these highly turbulent market conditions, would fall far short of meeting the profit goals that had been set as the measure of system effectiveness. G.M., therefore, had a problem—namely, to determine the best strategy to meet its goal of profit maximization.

G.M.'s actual reactions over those two years affected all aspects of the cybernetic systems diagram. On the *output* side of the system, G.M. tried a rebate program. The purpose of this strategy was to sell more rapidly the cars they already had produced, or by making these cars more attractive to consumers through in effect lowering the price.

Next they set a lower level of car production and thus changed both future systems output and the then-present *input*. Less need for resources such as materials and manpower could reduce costs. A result of this policy was the very large layoff of assembly-line workers.

A longer range strategy which G.M. had already started implementing was the switch to a greater small- versus large-car mix. In addition to affecting the system's output and input needs, this also entailed a change in *process*. New production plants and tooling requirements were needed to insure the success of this program.

On the *environmental* side of the business system, G.M. initiated a strategy designed to reduce the rapidity of change and lessen the demands from its environment by trying to get the Federal government to extend the time limits when certain auto safety and pollution emission requirements would become law. Lastly, G.M. could have changed its *measures* of *system effectiveness* by simply lowering its profit goals to better match the realities of today and tomorrow as opposed to yesterday when the goals were set.

This very brief example is used to illustrate how a business can be looked at as a cybernetic system. Furthermore, each of the strategies is really an alternative way of structuring the company system.

FURTHER ELABORATION OF ENVIRONMENT

As was defined earlier, everything outside of the system boundaries is considered the environment. The general forces in the environment of a cybernetic system are pulling or pushing the system in various directions, which may result in greater or lesser system performance. These environmental forces can be divided into an *immediate environment* and a *general environment* (Kotler).

The immediate environment includes those factors which, when they change, have a direct and immediate influence on the system and its performance. On the other hand, the general environment includes forces which are much more indirect in their effect on a system. While the general environment can have a tremendous impact on a system, these forces usually change magnitude and direction very slowly (Kast & Rosenzweig).

To give an example of these points, let's consider a business as a cybernetic system. The immediate environment would be the firm's suppliers, competitors, customers, and the direct impact of what we will call socio-political forces (see Fig. 3-7).

As part of the immediate environment, socio-political forces comprise elements like the government, the state of the economy, trade unions, consumerism, etc., which can have a direct and significant influence on how effective a business functions. The government can affect a business in many ways such as corporate taxes, anti-trust laws, price controls, monitoring of wage increases, purchasing, etc. Whether the general economy is in a recession, a stable period, or a growth condition affects businesses. Trade unions can pressure a business through the restriction of input resources and/or boycotts of business output. Customers of various businesses have been massing in number to acquire power to influence how and what a business produces.

The output of a business is very much dependent upon its means for getting the desired mix of resources in the form of men, machines, money, and materials. The ability of suppliers to have these resources at the time they are needed can have a significant impact on a business.

The customers of a business use its output as input to their system process. As the users' needs or desires change, their estimation of the value of the firm's product or service will change. Competitors of a business can also have a direct impact on how well the firm does. Customers of a firm can be lured away by competing organizations which offer a product or service which better meets the needs or desires of the users.

The general environment, on the other hand, can be looked at as the singular or combinational effect of political, cultural, technological, and resource-base forces. These forces are long range in scope and generally change very slowly. However, they are the underlying forces which eventually cause changes in the immediate environment.

Cultural forces include the values, ideologies, and norms of the society in which the firm is doing business. Political forces include what society's relationships are with other countries in the form of war or peace, isolation vs. internationalization, etc. Technological forces include the scientific and technological advancement of a society in terms of the knowhow and the type of "machines" it can produce. Resources is a concept which applies to both the natural and human base of a society and to show how its abundance or limitations can have a rippling effect on the immediate environment of a business.

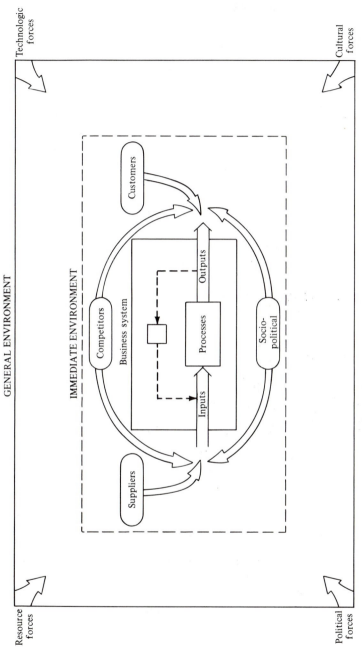

Figure 3-7 Environmental factors—business system.

35

INFORMATION SYSTEMS DEPARTMENT
AS A CYBERNETIC SYSTEM

In Fig. 3-8, the Information Systems Department at California State Polytechnic (Cal Poly) University is looked at as a cybernetic system. When first looking at the diagram it appears to be very busy and quite overwhelming. Taken one step at a time, however, we find the diagram depicts and summarizes all the major points that we have considered so far in Chapters 2 and 3.

Working from the center of the diagram, we see that the Information Systems (IS) Department is shown in the cybernetic format of inputs, process, outputs, and objectives, etc. The environment, however, is a little more complex than we have seen. First, the IS Department is a subsystem to the School of Business Administration which totally encompasses it in the diagram. Further, the School of Business is a subsystem to Cal Poly University, as are the Schools of Engineering, Arts, Agriculture, etc. But Cal Poly is not a stand-alone system; rather it is a part of the nineteen campuses belonging to the California State Universities and Colleges system. This organizational environment for the IS Department is clearly of a different type than the immediate and general environment we have previously defined.

The IS Department is itself a system since it is made up of components such as students, full-time faculty, part-time faculty, courses, office space, classrooms, budget, etc. In a cybernetic sense, the IS Department has inputs of students who are majors; students who take service courses; money, buildings, teachers, computer facilities, etc; and knowledge. Note where these inputs come from. For example, money, buildings, etc., come from the administration of Cal Poly University. This is a very critical interface for the IS Department in the attainment of its own objectives. Another critical interface is with the other departments (i.e., Marketing, Management, etc.) within the School of Business who require their students to take IS courses. In the case of the IS Department, approximately 2/3 of the student demand for courses comes from nonmajors! Again it is crucial to the IS Department to satisfy the CDM's of these nonmajors.

Who else are suppliers of the inputs to the IS system? The student majors can come from high schools, junior colleges, adult education, etc. But the reality is that over 50% of the IS majors at Cal Poly have first attended a community college! Hence there is a need for a smooth interface with the local community colleges and a real opportunity to tap the local high school to tell these students the advantages of coming directly to Cal Poly instead of attending a community college first.

In terms of outputs, the IS Department produces graduates with a degree in Information Systems. But this isn't the only output. Information system skills are given to nonmajors who take IS courses.

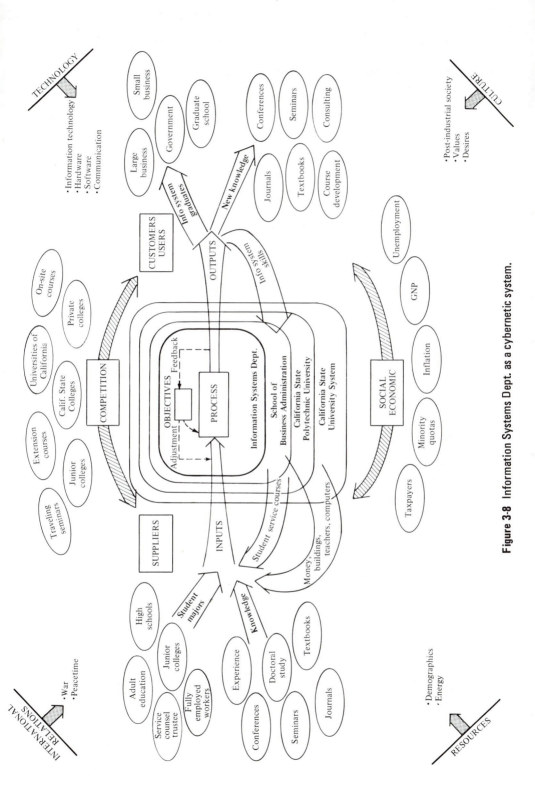

Figure 3-8 Information Systems Dept. as a cybernetic system.

37

Further, new knowledge is disseminated in the form of faculty speaking at conferences, giving seminars, writing textbooks, etc. Who are the users of the IS Department output? Perhaps the primary customers/ users are the eventual employers of IS graduates, that is, businesses, governmental agencies, graduate schools, etc. Here again, these are important interfaces because if your students have not been well educated, their chances for the good jobs and first-rate graduate schools will be slim. The graduates in turn will be displeased and will encourage their friends and relatives to attend a different college or try a different major, etc.

While universities don't like to talk about being in a competitive environment, they certainly are. In our basic diagram the competitors for students are community colleges, other California state universities, other University of California campuses, private colleges, etc. The outputs from these alternative sources (i.e., students educated in Information Systems skills) also compete for jobs, entry to graduate schools, etc.

In terms of the socio-economic impact of the immediate environment, consider the result of the taxpayer's revolt (i.e., Proposition 13 in California in 1978) on money for opening new campuses, the number of classes which can be scheduled on any one campus, the salary paid to faculty and staff, etc. What about the impact of Affirmative Action requirements on the university in terms of hiring faculty and staff?

Longer range considerations are shown by the factors of technology, culture, politics, and resources. In the case of the IS department, it is quite easy to anticipate a great impact of these factors on what the IS Department teaches, since computer-based technology has changed rapidly over the last ten years and even more changes are forecasted to take place over the next ten years. What about the impact of dwindling gasoline supplies on schools whose students are primarily commuters? What impact would the universities feel if the United States were to enter in another major war?

This brief look at the Information Systems Department as a cybernetic system was designed to give you some insight into the complexity of real-life systems problems. As analysts we will need to know how to cope with these types of problems. One concept which will help is that of scenarios.

SCENARIOS

To give an indication of how the concept of a scenario is used, recall the systems context diagram in Fig. 2-1, where we were studying the Jones family and their desire for a new car. The major environmental factors were listed as company, inflation, gasoline prices, air pollution, and relatives. All these factors in general affect the Jones family and the Joneses have no direct control over these forces.

As the Joneses study what is the best car for them, they need to consider what the price and availability of gasoline will be during the planning horizon.

> **Planning horizon** is the assumed time period for which any system solution must be effective in order to be acceptable.

In the case of a new car, the Joneses are assuming they will own it for five years. Therefore, the planning horizon is the five-year period 1980–85. Thus, the Joneses have to estimate the situation with gasoline for 1980–85. Why? Because it will affect their decision on what is the best car for the Jones family if the price of gasoline is $2.00 per gallon. How about the impact of a gasoline rationing scheme which limits a family to 30 gallons of gas per month and Mr. Jones drives 50 miles per day to work presently and he has no plans of either switching jobs or relocating their home?

Most people, unfortunately, either don't give any consideration to environmental factors or just assume implicitly that tomorrow will be like today. In a relatively static and unchanging environment, this approach is okay. But in our rapidly changing world of the 1980s, change in most sectors is very rapid. Therefore, the Joneses need to make a best estimate for the gasoline situation even though they realize that, in most cases with environmental factors, nobody knows exactly what the future holds.

After studying the situation, they decide their best estimate for the period 1980–85 is that gasoline will not be rationed, but will cost on the average $1.75 per gallon. Optimistically, the best that reasonably could be expected (i.e., not what you desire) is no gasoline rationing and a price of $1.25 per gallon. Pessimistically, the worst that reasonably could be expected (i.e., not what you dread) is gasoline rationing of 30 gallons per month per family and a $2.50 per gallon price.

In summary, then, for the Jones family during 1980–85 the scenario factor is gasoline with the following estimates:

Most Likely: No rationing and $1.50 per gallon

Optimistic: No rationing and $1.25 per gallon

Pessimistic: Rationing of 30 gallons per month per family and $2.50 per gallon

We have just looked at one factor, what about the others? The Joneses' decision on buying a car may not be affected by the status of relatives (i.e., parents or parents-in-law) over the 1980—85 period. However, a change in his job could greatly affect both his income and where his family lives, which in turn should affect his car-buying decision. Therefore, the Joneses' scenario should include gasoline prices and job status, but not include relatives.

Only those factors which are forecasted to change and if they did change would have a major impact on the decision at hand should be included in the scenario. For very short planning horizons, the scenario will just contain factors from the immediate environment, because the general environmental factors will not have had time to make an impact on the decision. However, as the time frame is lengthened the scenario becomes predominately the consideration of general environmental forces as they will impact the system under study.

In the Jones family example, the scenario factors were relatively straightforward and we made no differentiation of immediate and general environmental factors. But to see these points, let's reconsider Fig. 3-8, the Information Systems Department as a cybernetic system. In studying this diagram, it is very clear that one couldn't possibly consider all the environmental factors which if they changed could influence the IS Department. Therefore, the analyst has to be selective and analyze the most important forces as they affect the decision under study. Another approach in any complex situation such as this is to include many factors but just consider the most likely outcomes, not the range from optimistic to pessimistic.

Stated in this way a scenario for the IS Department during the 1980 to 1990 time period could be:

Demographics:	The number of 18—24 year old potential students decreasing by 15%
Politics:	Draft of college-age men and women to support military
Culture:	Sustained demand for Women's Liberation Movement in terms of meaningful careers for women college graduates
Job Outlook:	A 15% increase per year in job demands for Application Programmers and Analysts
Faculty:	Teachers' pay will continue to lag behind cost of living increases and will average a 5% per year disparity

With these examples as background, we will define scenarios as follows:

Scenario is a collection of estimates of the future status of major system contextual factors.

Class Exercise 3-1

INFORMATION SYSTEMS DEPARTMENT

Taking Figure 3-8, the Information Systems Department as a cybernetic system as a base along with the previous explanation, discuss the following points:

1. What impact has women's liberation had on the University? What might the future effects be?
2. What have or could universities do to counter the impact on them of the trend that the number of people 18 to 24 years of age will decrease by 15% over the next ten years?
3. What is the process the IS Department (or any other department) has?
4. What are or should be the objectives of a college department? Is it to maximize profit? Is it best job placement for students? Is it educating the whole person? Is it security for the tenured faculty?
5. Can a department have too many student majors? What is the "right" number?
6. What are some of the major interfaces for the IS Department which should be closely monitored and managed? Are some interfaces more important than others?

Class Exercise 3-2

INDIVIDUAL AS A SYSTEM

1. Can an individual be considered a system?
2. How would a person act if he were functioning as a (a) static system? (b) dynamic system? (c) homeostatic system? (d) cybernetic system?
3. Comment on the statement, "No man is an island." Relate your system thoughts in this regard to question 2.
4. If you were functioning as a cybernetic system, what are your system objectives? Relate this to Maslow's hierarchy of needs. How do fixed vs. adaptive goals fit this discussion?
5. What are some immediate and general environmental forces acting upon a college student today?
6. What are some potential career system designs for you? Comment on the statement, "You can be anything you want to be—if you try hard enough."

Class Exercise 3-3

SCENARIOS

The airline and home building industries are two which are highly affected by the business environment.

1. The airline passenger business is comprised primarily of business and consumer vacation travelers. Look at the relationship of airline profits and the health of the economy during the 1970s. What happened during the recessionary periods? What happened during periods of economic growth? Develop a scenario for airline planners to use for the 1985-1995 time period considering the health of the economy and the price of airline fuel.

2. What impact does mortgage rates have on new home sales? How is this related to the fluctuations in the prime rate? What is the impact of inflation on the prime rate? Give an optimistic, most likely, and pessimistic estimate for mortgage rates for new houses during the 1983-1988 planning horizon.

Chapter Four

Problem Formulation

Basic to the attainment of a solution or solutions to a problem is an understanding of what really constitutes a problem. Basically all problems are deviations from what is expected for a particular system in comparison to what is actually happening. Or stated in other terms, what we want compared to what we have. In this chapter we will define the different types of problems and relate them to the more common operational, tactical, and strategic level decisions.

Further, we will see how to establish what the problem really is, first by doing a problem diagnosis of the system, which will result in establishing symptoms, which in turn will help define the problem cause, and will finally lead to the problem identification. Lastly, the ways to cope with problem complexity when dealing with real-life systems will be discussed.

TYPES OF SYSTEMS PROBLEMS

While all problems are deviations, there are three different ways this difference can occur. The first type may be called a *negative deviation*.

Assume in Fig. 4-1 that horizontal distance measures performance levels for a system. Further, distances to the right of the page indicate increased performance, and movement to the left of the page shows decreased performance.

According to our definition of a problem, a time frame must also be considered. Let t_0 indicate performance as of a particular point in time. Let t_i indicate a future time period measured in hours or days or years. The planning horizon will be shown as a certain period of time like $t_1, t_2, \ldots, t_n$. Further, we define an outline of a triangle, $\triangle$, to represent performance that is expected and a filled-in triangle, $\blacktriangle$, to represent performance that has or will actually take place. Now we are ready to proceed to the different types of problems.

Negative Deviation

In Fig. 4-1, assume we expected a certain performance level at time t_0, t_1, t_2, up through t_n. This is shown as $\triangle t_0, t_1, \ldots, t_n$. The system is presently performing at this same level (i.e., $\blacktriangle t_0$). So, by definition, there is no problem.

However, assume at time period t_1, that the systems performance decreases to a level depicted by $\blacktriangle t_1$ in Fig. 4-2. Since there is a deviation between the performance level we expect or desire, and what we are now getting, we have a problem. Further, note that the deviation is in the negative direction.

An example of this type of problem is when you go out to your car to drive home after work and the car won't start. You expected to get in your car and drive home, but possibly due to a dead battery, your car will not perform like you have come to expect.

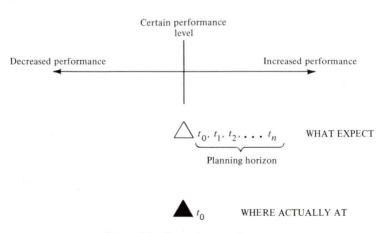

Certain performance level

Decreased performance   Increased performance

$\triangle$ $t_0, t_1, t_2, \ldots t_n$ WHAT EXPECT

Planning horizon

$\blacktriangle$ t_0 WHERE ACTUALLY AT

Figure 4-1   No problem at time t_0.

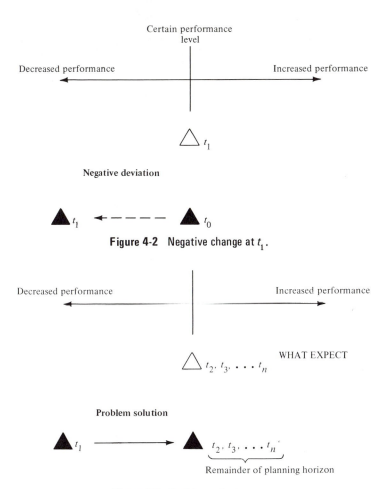

Figure 4-2 Negative change at t_1.

Figure 4-3 Problem solution at t_2.

Generally, with this type of problem, you try to fix the system so that you get the actual performance level back up to where it was. This is shown as ▲ t_1 ⟶ ▲ $t_2, \ldots, t_n$ in Fig. 4-3. That is, you could recharge the battery and continue on your way home. Your intent in solving this type of problem is not to improve the overall system per se, but rather to just repair the situation. This type of problem is summarized in Fig. 4-4.

Positive Deviation

In Fig. 4-5, we again start at time t_0, where the expected performance △ t_0 of the system is the same as the actual performance ▲ t_0. Assume the system will continue to perform at this level in future time per-

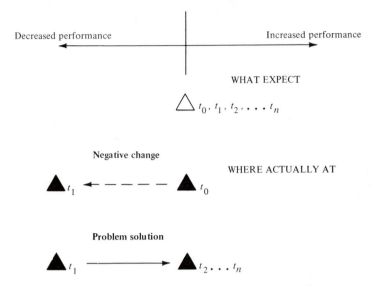

Figure 4-4 Summary chart of negative deviation problem.

iods. So we shouldn't have any negative deviation type problems.

But now assume the CDM for this system increases his or her expectations of what can or should be desired from this system to a level indicated by $\triangle\, t_1$. Because there is now a deviation between $\triangle\, t_1$ and $\blacktriangle\, t_1$, there is a problem. This type of problem indicates a positive deviation.

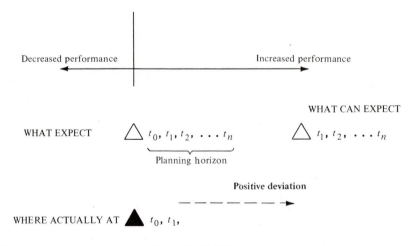

Figure 4-5 Positive change.

The approach for solving this type of problem is to determine how one can increase the actual performance of the overall system. For example, with many computer solutions now available to organizations, performance levels that were acceptable five years ago are no longer satisfactory. A possible solution then, might be to improve the organization by purchasing one of the latest computers. An overall summary is shown in Fig. 4-6. Be sure to note that there has been no decrease in performance of the system. It is working as it has been designed. So, per se, there is no problem (i.e., operational level). The problem is the result of new awareness of what can be reasonably expected from this system.

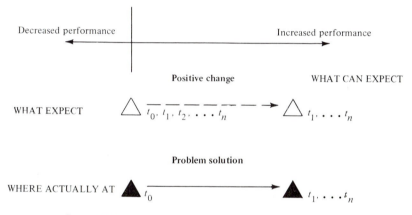

Figure 4-6 Summary chart of positive deviation problems.

Fundamental Deviation

In the third type of systems problems, we again start where the expected level of performance $\triangle t_0$ and the actual performance $\blacktriangle t_0$ are the same. But now, possibly due to a significant change in the environment in which the system is embedded or a change in the underlying values of how systems performance should be measured, there is a fundamental change in the way a system should be operating.

For example, according to Gail Sheehy's best-selling book *Passages*, there are several crisis points in a person's life. One such point is around 30 years of age when most people come to realize that the path they have set for themselves coming out of college is not necessarily the best for them anymore. Different values become less or more important. Perhaps money or prestige doesn't look as important as personal freedom. Or conversely, power and status could become lots more important than leisure time. This in turn should affect the implicit values that one uses to measure personal effectiveness in life.

In the above example, nothing was said about the environment. It was seen as constant. Assume now a person's life values are held constant and we significantly change the environment. The person is quite satisfied with job, living situation, etc. He or she has an excellent job as, say, an elementary school teacher. However, the job market for teachers is contracting and the future looks grim. This change is due to the reduction in the birth rate which in turn is caused by a change in cultural values, technology, etc. (i.e., women's liberation, abortion, the pill, etc.). Thus, even though the person is doing a good job and loves the work, the environment has significantly changed and thus forces a fundamental change in the person's "system" which has very serious and long-term consequences.

Two highly different reactions are possible in this type of situation. The person can view the environmental shift as a threat, and can adapt to the situation by lowering his or her expectations of what he can accomplish in life. The other reaction is to take the shift in the present environment as an opportunity. In this example, the person could enroll in night school in order to pursue another career.

To see a pictorial representation of these fundamental deviations due to (1) a significant shift in the environment and/or (2) an important change in underlying systems values, consider Fig. 4-7. In this diagram a shift in values of how to measure performance is shown as a tilled line ↗ , otherwise the notation is the same as before.

In assessing the future environment, assume it is seen that this situation is not favorable and/or there has been a shift in values which leads to a more restricting situation. This combination is shown as a *threat deviation* in Fig. 4-7 and the problem solution is to reduce expectations $\triangle t_1, t_2, \ldots, t_n$ to the actual level attainable in this new environment $\blacktriangle t_1, t_2, \ldots, t_n$.

Conversely, if the future environment is very supportive of positive change and/or the shift in values leads to more meaningful performance, we label this an *opportunity deviation*. The problem solution is to find a solution which will increase the actual and expected performance to the right and upward.

RELATIONSHIP OF PROBLEMS AND DECISION LEVELS

In the previous section, we developed a means of classifying the three basic types of systems problems. Here we want to clarify this relationship with that of the level of decision required by each problem.

Within any organizational system (e.g., business, church, family, individual, United Nations) there are basically three levels of decisions—*strategic, tactical, operational* (Anthony, 1965).

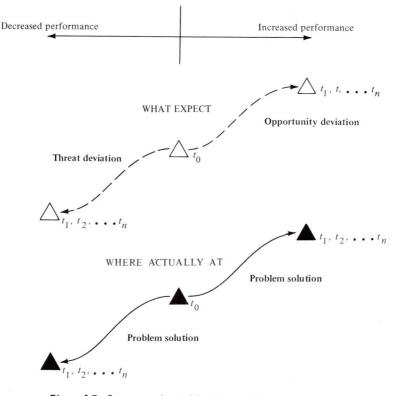

Figure 4-7 Summary chart of fundamental deviation.

Strategic decisions are those decisions which involve the determination of what the overall system objectives should be. It basically tries to answer the question of what the purpose of the system is. *Where should the system be going?*

Tactical decisions are those decisions which involve the determination of how to best accomplish the overall system objectives. *How do we do it better?*

Operational decisions are those decisions which involve the carrying out of the system objectives while keeping the system within constraint limits. *Doing it.*

To see how these various decisions relate to a system, let's consider the German company that makes Volkswagens.

The strategic decisions for this business system would be the deciding of what the purpose of this organization should be and what kind of business it should be in. For VW this could be establishing that the organization's purpose is to maximize profit within the auto industry.

Tactical decisions would involve establishing how to maximize profit within the auto industry. The result of these decisions could be to sell a standardized subcompact car in the United States market.

Operational decisions would involve the carrying out of the tactical decisions by producing the cars, figuring out the best way to transport cars to the USA, watching the cash flow so that it does not drop below constraint limits, etc.

Table 4-1 shows the overall relationship between the type of systems problems and the level of decisions required. It is important to recognize that with every system (subsystem), there is a need at various times to make decisions regarding all three types of problems.

For example, assume the system is an academic department within a university. For the department, operational decisions would be developing schedules on which courses are offered and at what times by which instructors. Further examples, after the semester has started, would be ensuring that there are enough students, rooms are in fact available, etc. (i.e., Doing it).

Tactical decisions would include redesigning course content and developing more effective curriculum design; That is, assessing how good the present curriculum is and trying to come up with even better designs (i.e., How to do it better).

Strategic decisions involve conjecturing how the environment will change for graduates of this program over the next five to twenty years; assessing what type of orientation faculty should have regarding research and teaching; establishing what concentration should be devoted to graduate versus undergraduate studies; determining whether a new Master's program should be offered in EDP Auditing versus the more traditional M.S. in Information Systems, etc. (i.e., What should we be doing?).

Table 4-1 Relationship of problems and decision levels.

Type Problem	Level of Decision	Description	General Statement
Negative Deviation	Operational	Maintenance	*Doing it* Keep it running
Positive Deviation	Tactical	Management Planning	*How to do it better* Better benefit/cost ratio
Fundamental Deviation	Strategic	Policy	*What should we be doing?* Redirecting system

PROBLEM DIAGNOSIS

In the first section of this chapter problems were defined as deviations between expected and actual systems performance. But what was not answered was how one recognizes when this situation has or will occur.

Symptoms

The usual way of recognizing that something isn't as it should be or as it is expected to be is through the concept of symptoms.

> **Symptoms** are indications that the system isn't performing as expected or desired.

That is, we compare along certain dimensions what we expect and what we actually see or measure, and these deviations are defined collectively as *symptoms*. For example, a mother takes her child to the doctor when the child shows indications which are not normal or usual. The doctor asks what's wrong, or what are the child's symptoms? The mother responds that the child has a fever, running nose, is cranky, and has been sick to his stomach. From these indications the doctor attempts to determine the cause. In this example, it could be the flu, or something the child ate, or a reaction to a shot, etc.

Cause

In our example of the sick child, the mother doesn't necessarily know what is causing the problem—only that the child isn't acting like he usually does. But symptoms are neither the cause nor the problem!

> **Cause** is the fundamental thing which is generating the problem with a particular system.

The doctor tries to establish from the various symptoms what is the most likely cause, in this case the disease that is attacking the system and that results in the body giving off the particular set of indications.

Problem Identification

But it is important to recognize that in problem-solving, one shouldn't automatically assume that the problem to be solved is simply to remove the cause. Rather, problem-solving asks the question, What should be done to make the overall system most effective now and in the foreseeable future?

> **Problem identification** is the establishment of what question should be addressed in making a system more effective.

With the definition given above for problem identification, one can infer that identifying the problem to be solved is neither straightforward nor easy. How the problem is defined heavily influences the analyst's perceptions and the potential solution that he would consider. According to the old saw, "A problem well defined is half-way solved."

Potential Solutions

With the identification of the problem, one can then more easily establish what should be done.

> **Potential solutions** are methods which seem like they will solve the problem that has been identified with a particular system.

Let's use an example from the realm of house plants to show how these concepts of symptoms, cause, problem identification, and potential solutions all relate (see Table 4-2 (a) and (b)).

The difference between the two charts is important for our future understanding of how to deal with much more complex systems. The symptoms specified are three separate indications as to behavior which is abnormal for these kinds of plants. These may appear separately or in combinations, which could imply different potential solutions.

The cause is stated the same. However, in the first-aid chart there is no explicit statement for problem identification because it is assumed that the problem is to remove the cause. Generally not a bad assumption, but tends to focus the analysis on the wrong thrust.

The what to do section has been retitled potential solutions to indicate these are methods that will probably work (i.e., solve the problem stated of too much water). But even here, they are three separate solutions depending on what the actual cause is.

To see why problem identification is important and how it can influence the perception of the situation and the search for solutions, consider what the alternatives might be if we defined the problem as: What should be done about the sick house plant? One of the alternatives could be to water less frequently, another approach could be to throw the plant out. A third solution could be to cut off sagging leaves, put bright ribbon on the clay pot, and then give plant as a house warm-

Table 4-2 (a)

House Plant First-aid*

Symptom	Cause	What To Do
Lower leaves turn yellow; stems become soft and dark in color; green scum forms on clay pots.	Too much water	Make sure the pot's drainage hole is not clogged. If the soil has become compacted, roots may decay for lack of oxygen; repot the plant. Water only if necessary.

*Extracted from chart in *Foliage House Plants*, Time-Life Encyclopedia of Gardening, 1976.

Let's take this chart and rework it into our scheme.

Table 4-2 (b)

House Plant Problem Diagnosis

Symptoms	Cause	Problem Identification	Potential Solutions
(1) Lower leaves turned yellow	Too much water to the plant	What is best way to get and keep plant in healthy state?	(1) If drainage hole is clogged, clean out
(2) Stems are soft and dark in color			(2) If soil compacted, repot the plant
(3) Green scum has formed on clay pots			(3) Water less frequently. Water 3 times a week, 2 oz. each time

ing gift to your new neighbors. A fourth approach is to learn to like yellow leaves!

PROBLEM DIAGNOSIS AND DECISION LEVEL EXAMPLES

Let's carry these concepts to much more complex systems examples and further relate them to the level of decision.

In the first example, Table 4-3, the symptoms are water spots on

Table 4-3 Family-oriented example.

	Symptoms	Cause	Problem Identification	Potential Solutions
	Water spots on living room floor	Hole in roof It is raining	What is best way to handle water spotting?	Buckets Patch roof hole Refinish total roof
Operational				
	Frustrated Grouchy	Hot, smoggy summers	What's best way for you to handle heat?	Get used to sweating Beer Get air conditioner Swimming pool
	Lots of tension in house among kids during morning and evening	New baby 3 kids in a room Too crowded	What's best way to get more room? or What's best way to help kids get along better?	Sell baby Add on room Move to larger house/ same neighborhood
Tactical				
		Property tax has doubled Pay has not kept pace	What is best allocation of money?	Set up budget Rework priorities Move to apartment

Table 4-3 Family-oriented example (continued).

	Symptom	Cause / Desire	Question	Options
Tactical (cont.)	Strapped to meet monthly house payment	or Other money demands from family needs or	How to get more income?	Get part-time income Encourage wife to work Change jobs Take out 2nd mortgage
		Husband sick, can't work	Best way to get husband well?	Operation in hospital
	Headaches Ulcer Passed over for promotion	Desire for better life		Stay in L.A. Work downtown Advance to top or
Strategic	Kids not knowing or caring about nature Lost enthusiasm	Hearing about the happiness of the Jones family	What is best life style for Brown family?	Move to Oregon Open sporting goods store Do part-time farming or Move to Australia Live in commune Work for church Live each day to fullest

the living room floor. When we say they are symptoms, we are implying this isn't the usual situation. Further, some processing had to be done to establish that the spots are in fact water as opposed to oil, blood, urine, etc. The clearer indication we have of the symptoms, the easier it will be to determine the cause. In this case, the cause could be the rain coming through the roof. The problem identified could be to keep rain from coming through the hole in the roof, or the focus could be on the best way to handle water spotting in the future.

If the problem is identified as keep rain from coming through the roof, then solutions like patching the roof are reasonable. But what if the family can't afford this expense or the rainy season is over and they are planning on moving next month anyway? Or what if their overall plans are to upgrade the whole house? Depending on how the problem is stated, the analyst would focus on different solutions. Note the solution of using buckets. Sometimes living with a situation is the best solution! It is not always desirable to fix or remove the cause of the problems. In any case, this type of problem requires an operational decision.

In the business-oriented example (Table 4-4), the tactical level decision shows various symptoms to include user's dissatisfaction with the usefulness of reports and their accuracy and timeliness. Further, to process the computerized reports, the data processing department has had to go to working on weekends and the employees are upset with this arrangement.

What's the cause? The function of salesmen is to make you dissatisfied with your present situation. By our definition of a problem, if you are not dissatisfied there is no deviation between what you expect and what you have. Therefore, there is no problem. Thus, you aren't going to buy (i.e., change alternatives). Why should you?

The salesman could be a means for providing you information about a new computer technological breakthrough which will enable your company to develop an in-house computer-based capability. In so doing, management can and should realize their expectations as to what information reports they can receive.

The problem identification statement focuses on the question of the best way for the company to provide information from its accounting system. One of the alternatives is to do nothing. That is, stay with the present system. In tactical level decisions this makes sense because there aren't any repair problems needed to keep the system viable. Other alternatives include adopting a prepackaged system and developing a custom information system in-house.

As you work through the other parts of these examples, concentrate on the interconnection and flow of symptoms, cause, problem identification, and potential solutions.

PROBLEM COMPLEXITY

So far in this chapter we have discussed systems problems, how they can be diagnosed, and how this relates to decision level. Nothing has been said about problem complexity and how to cope with it.

> **Complexity** is a function of the number and types of elements and their degree of interaction.

With this definition, a pencil would not be a very complex recording instrument. Whereas, a typewriter, both because it has more elements and the degree of interaction is more varied, would be a complex recording instrument. A computer exemplifies great complexity.

Factors Influencing Problem Complexity

One of the major factors influencing problem complexity is the inherent complexity of the system being studied.

Systems as we had stated are sets of components which could be looked at as working together for one or more overall objectives. Through the concepts of the systems context diagram and the basic cybernetic system (Fig. 3-5), we have begun to get some feel for the complexity surrounding the studying of systems. We now need a more formalized definition of systems complexity.

> **Systems complexity** is a function of the number and type of components and the level and degree of their interaction.

Relatively simple systems are comprised of very few components having a limited number of attributes, where the interactions among components is either very limited or extremely well patterned and predictable (Schoderbek, et. al.).

A chair is an example of a very simple system. It is generally composed of components such as a back, legs, seat, frame, etc., whose interaction is singular and fixed. On the other hand, a hospital is an example of a very complex system. The components would include patients, nurses, doctors, administrators, machines, medicines, etc. These factors interact in a highly sophisticated manner when an experienced surgical team performs a crucial operation on a patient who has just had a heart attack.

Appreciating the underlying complexity of a system will affect the specific strategies used to increase the effectiveness of the system. The simpler the system, the more likely we can determine which system de-

Table 4-4 Business-oriented example.

	Symptoms	Cause	Problem Identification	Potential Solutions
Operational	Cash balance below 1 million dollars	Cash flow is cyclical or Lack of profitable business opportunity	What is best way for Lockheed to keep enough cash in reserve?	Tighter cost controls Sell off assets Ask customers for prepayment Ask bank to reduce required balance
Tactical	Dissatisfaction with accuracy or timeliness of reports Users complaints regarding lack of certain management information Employees unwillingness to work on weekends	Salesman's message of increasing expectations New computer technology breakthrough Developing in-house computer-based capability	What is best way for company to handle its accounting system?	Stay with present system Adopt prepackaged computerized accounting system Develop own in-house computerized accounting and financial information system

Table 4-4 Business-oriented example (continued).

Symptoms	Cause	Problem Identification	Potential Solutions
Losing money return on investment	Competition with air-lines	What business should Penn Central be in?	Lower expectation reduce to minimum rail operations
Increased difficulty in hiring young, sharp college graduates	and/or	or	Get out of railroad business; switch to air freight
Sharp decrease in passenger and freight business	Businesses in Northeast relocating to South	What should be done with Penn Central?	Switch to real estate holding company
			Combination of above
			Turn over to federal government

Strategic

signs will work best. Extremely complex systems like the ecological relationships of the earth are so difficult to analyze as to boggle the mind. Much lower degrees of confidence can be attached to the effects of various changes to these types of systems.

The relationship between the complexity of the systems problem being studied and the underlying system is not one for one. In other words, the problem studied is an abstraction of the real situation.

> **Abstraction** is the degree to which something differs in representation from the actual situation.

Ideally, there should not be any need for abstraction, and one would work with the actual system. Unfortunately, because of the complexity of most systems problems this ideal is not attainable and some form of abstraction is necessary. The key is to make the abstract problem respond very closely to the critical aspects of the real system.

> **Problem complexity** is a function of the degree of abstraction of the system and its inherent complexity.

We are using the terms abstraction and modeling to be synonymous. A model, whether it be a mental model, a computer/mathematical model, a wooden model, etc., is a representation of some aspect of reality. The more complete the model, the less abstraction, and thus the greater the problem complexity. The more complex the system being studied, the more complex the problem.

Reducing Problem Complexity

Most generally, the inherent complexity involved in truly understanding many of the systems we are addressing in this text (i.e., business, governmental agencies, university, families, individuals, etc.) is well beyond man's capabilities. This is not to suggest giving up hope, rather it is an explicit statement to help insure that problem solvers realize they need to deal with situations involving uncertainty. We now will discuss ways which have proven useful as means for limiting complexity in effective ways.

Appropriate Systems Level: In Chapter 2 we considered the example of trying to improve the reading skills of first graders in Midtown, USA. This was a subsystem to all the skills first graders were being taught. Additionally, the first grade was a subsystem of that elementary school, which in turn was a subsystem to the school district, etc., all the way up to this as a subsystem to the world. It is clear we can't handle all these facets. To reduce the complexity to a more manageable form, without sacrificing much generality we defined the system in terms of who was the chief decision-maker (CDM) for the study. If the CDM was the elementary school principal, then all supra-systems such as the school district, etc., would be considered the environment. Through the concepts of scenarios (chapter 3) and external restrictions (chapter 6), we can consider the most important aspects of this particular environment. This decoupling will greatly reduce the overall systems complexity.

Critical Systems Components: In the systems context diagram (Fig. 2-1), the major components of the Jones family were represented. If the Joneses were thinking about getting a new car, the appropriate systems level *is not* the car. It is the Jones family, as defined by the control displayed by the CDM (i.e., Mr. and Mrs. Jones). But one doesn't have to consider all the actions and interactions of the Jones family subsystems to make a good choice of a car for them. In the interest of reducing complexity, many of the subsystems can be held constant and assumed to have no interaction which significantly affects *this decision.*

The Jones family components of furniture, food, and clothing have no direct interaction with the car decision. An indirect interaction could be assumed that money for the car is money that can't be spent for new furniture. If the family situation is relatively stable, then one can assume that Mr. and Mrs. Jones will stay married over the planning horizon and additionally that they don't plan on having any more children. The fact of whether the Joneses are a one- or a two-car family could affect which components must be considered. For example, if the car decision is for the second car for Mr. Jones to use in driving to and from his job, then the major systems components are Mr. Jones, his job, the family income, and possibly their life style. However, if the car decision is for the first car, then the major factors are the husband, wife, children, cash, job, recreation, pet, life style, etc., as it affects the primary uses of the car, i.e., shopping, church, Little League, family vacation, etc.

In the sense of systems complexity, the first-car decision would be more complex than the second-car decision because more system components must be considered as to their status and future interaction.

Planning Horizon: As was noted in chapter 3, the planning horizon is the assumed time period for which the solution is expected to be effective. The longer the planning horizon the greater the systems complexity. This should make sense since the greater the time period, the more that things can change. Thus, the need to preconsider more elements of the system prior to making the decision.

In the case of selecting a mate, if you believed in marriage and specifically the vow of "until death do us part," what is your planning horizon? Assuming a couple got married at age 20 and had a life expectancy to age 70, then the implicit planning horizon is fifty years! No wonder very few marriages fulfill this pattern. The amount of change that takes place in a culture (environment) in fifty years can be astounding. In addition, the changes in values, interests, health, etc., of the marriage partners themselves is often greater. So it is clear that we can't hope to completely deal with the great complexity in selecting a mate over a planning horizon of fifty years. However, we can make some very good choices if we concentrate on say the first five years. This would reduce the systems complexity considerably, but would still get at the very crucial initial period.

The complexities and complications of the 1970s (e.g., Vietnam War, oil, nuclear reactors, inflation, etc.) have convinced most longrange planners that businesses, governmental agencies, etc., should be very happy if they could look out and plan effectively over a three- to five-year period. In general, the shorter the planning horizon, the less is the systems complexity because the components and their interaction are less changeable (i.e., more stable and predictable).

Problem Importance: How important is this problem to the overall system? What would the impact be on the system if the problem were ignored? That is, suppose we did nothing or continued to have the system function as it is presently. If there would be major questions of the system continuing to be viable or if there would be major disruptions in a negative sense, then the problem should be considered important.

In general, strategic level decisions are much more important than those on the operational level since they impact on many more of the subsystems and in such highly influential ways. Furthermore, strategic decisions by their nature tend to have an impact over a much longer time period than do either tactical or operational decisions. On the other hand, operational decisions impact very quickly on the viability of a system. If a firm can't meet its accounts payable requirements today because of cash flow problems, it isn't going to be around tomorrow to take advantage of a growth situation in the future.

Another consideration is how much difference is there in system effectiveness with the various potential solutions? The greater the difference in potential benefits derived from the solutions, the more time should (could) be spent studying the system.

What are the consequences of being wrong? If the situation is such that management could try one solution and if it didn't work try another without paying too high a price, then it probably isn't worth taking much time to study the situation. In other words, how much of your resources are you having to commit to this decision? How easy would it be to bail out if you are wrong?

Given this type of reasoning, the analyst can assess the importance of the solving the systems problem. The less important the problem, the less time the analyst should spend trying to determine the "perfect" solution. Many things can be assumed and in the extreme case of a very unimportant problem, the analyst could (should) recommend to let things happen as they may. This in turn is a way of greatly reducing problem complexity. Note, this is only effective with relatively unimportant problems.

Let's take an example of a family as a system. What kind of food to buy at the grocery store is a continual problem. However, how important is this question really? How much time should you spend on determining which is the "best" laundry soap? You need to ask questions like—how important are "white" clothes to you? Have you been able to see much difference in your clothes when laundering with soap A, soap B, or soap C? What are the consequences of picking the "wrong" soap? How easy is it to switch to a new solution? How difficult is it to determine, prior to making the decision, the performance levels of the various alternatives?

For most of us, washing is not that important a problem, therefore, we find a soap that works and continue to use it (i.e., brand loyalty). If something better seems to come along, we try it; if it is better we switch, otherwise we stay with our original solution.

GENERAL CHARACTERISTICS OF SYSTEMS PROBLEMS

In this chapter we have defined and discussed the various types of systems problems, classified as operational, tactical, and strategic. These concepts are summarized in the chart, Table 4-5.

The systems problem classifications are related to the types of

Table 4-5 General Characteristics of systems problems.

	Operational	*Tactical*	*Strategic*
Type Problem	Negative deviation	Positive deviation	Fundamental deviation
Planning Horizon	Present	Near future	Long-range future
Environmental Impact	Little	Moderate	Significant
Purpose	Viability	Efficiency	Effectiveness
Scope	Narrow	Wide	Extensive

problems which were defined as negative deviation, positive deviation, and fundamental deviation. With operational problems there is immediate concern because the planning horizon is today and the impact of the environment is slight.

With tactical problems, there is a need to consider the near future and assume moderate environmental impact. Strategic problems are basically long-range considerations in which the impact of the environment is very critical to the effectiveness of any solutions that are developed.

In chapter 10, we will discuss in detail how to do a system study. Two of the points that are of interest here are the study purpose and the scope of the investigation. In the study of operational problems, the major purpose is to restore the system's viability. The scope of the study is quite narrow. With tactical problems, the scope of the study is wider and the major concern is to increase the efficiency (i.e., benefit/cost ratio) of the system. Strategic problems by their nature require the scope of the study to be extensive. The purpose of the study in this case is to increase the effectiveness of the overall system.

Class Exercise 4-1

EDUCATION-ORIENTED SITUATION

In this exercise, refer to Table 4-6; assume you are a counselor who is working with students at a local college.

1. In talking with the first student you and she have established that she has an operational level problem. She has flunked her test and is behind in her term paper for a particular class. She wants to know what is the best way to pass this course, given her present situation. You have suggested she could attend class more, since she has missed several sessions, or she could study with a friend to improve her understanding of subject matter, etc.
   (a) Given those symptoms, problem identification, and potential solutions, what is the likely cause(s) for this problem?
   (b) How do you know if you have found the "right" cause?

2. With another student you have established that he has a tactical level decision to make. You have determined he has a GPA below a 2.0 and has continually dropped courses. The student has stated to you that he can't stand to study. After talking with him you feel the cause of his problem could be that he is taking too many units, or he is in the wrong major, or his family problems are overwhelming him at this time.
   (a) Can there be more than one cause to a problem?
   (b) What does it imply when the causes listed are at different systems levels?
   (c) Develop a problem identification statement which relates to one of the causes mentioned and which matches the potential solutions given.

3. In talking with Sally Brown, she has a much more encompassing strategic level situation. Her symptoms and problem causes have led you both to identify her problem as "What is the best career for Sally Brown?"
   (a) Given this background, identify three potential solutions.
   (b) What makes this a strategic problem as opposed to an operational one?

4. Is it possible to have potential solutions to a problem, before you know what the cause is?
   (a) Is this good systems thinking?
   (b) Can brainstorming about solutions help clarify what the problem really is?

Class Exercise 4-2

TYPES OF BUSINESS PROBLEMS

In the following examples of business problems, state why you feel there is a problem (i.e., expected performance vs. actual performance); whether this deviation is primarily negative, positive, or fundamental; and lastly, establish if the situation is operational, tactical, or strategic in nature.

1. In the mid-1970s, the Lockheed Aircraft Company was given a loan guarantee by the Federal government in an attempt to bail this company out of financial difficulties. Lockheed commercial and military aircraft were considered to be first-rate technologically and appeared to have a bright future in their markets, but the company was having serious cash-flow problems.

2. One of the major reasons for the tremendous market in small business computers, is that businessmen can avoid the cost of hiring additional people to keep up with the paperwork functions of their expanding business. Thus while they must increase their present expenditures, they can offset future salary costs in addition to doing many of the same accounting functions faster and more accurately.

3. Penn Central has long been one of the major corporations of this country. However, while this firm has been operating passenger and freight railroad business for some 100 years, many of the heavy industries of the Northeast have relocated to the South or drastically cut their operations. Further, since World War II the passenger business hasn't been able to compete effectively with the car, airplane, and in some cases the bus. Penn Central management wanted to decrease its railroad operations and get into the air freight business; and also develop their extensive land holdings into industrial parks and residential sites.

Table 4-6

	Symptoms	Cause	Problem Identification	Potential Solutions
Operational	Flunked test Behind in term paper		What is best way to pass this class (or keep from flunking)?	Attend class more Study with a friend See instructor during office hours for additional help Change sections
Tactical	G.P.A. below 2.0 Can't stand to study Continually dropping classes	Taking too many units while working or Family problems or Possibly in wrong major		Reduce units and re-schedule tough classes Switch to an evening job with weekends off Develop more self discipline Change majors

Table 4-6 (continued)

	Symptoms	Cause	Problem Identification	Potential Solutions
	Dissatisfied with future job prospects	Major change in economic outlook for liberal arts majors	What is best career for Sally Brown?	
	Lack of interest in major area	No real experience in field, prior to making choice of major		
Strategic	New desire to go to graduate school	Excited about trying oneself on greater challenges		

Chapter Five

Developing
System Solutions

In the previous chapter we learned how to develop a diagnosis of a systems problem to identify what in fact the problem is. In this chapter, we want to address how to devise solutions for these types of problems.

To do this we need to first understand what the terms system design, alternative, and solution mean. These terms are generally considered to be interchangeable, but we need at this point to distinguish their differences.

> **Solutions** are ways or methods for solving the system problem that has been identified.

As we previously discussed, the word "solve" is used in a continuous sense as opposed to solved/not solved. Generally, there are various methods that will solve a particular system problem. These various ways are designated as alternatives.

> **Alternatives** are the different courses of action or approaches which could be used to solve a particular systems problem.

When speaking of different ways, approaches, or courses of action we are speaking of changing the underlying system in some way. This could be changing or adding components, restructuring the relationship between subsystems, etc. These various options are thus system designs.

> **System designs** are concerned with the appropriate selection of system components and their arrangement (structure) so as to meet the overall objectives of the system.

Thus system designs are considered to be alternatives. After study, those alternatives which can be shown to solve the stated systems problem will be considered solutions.

WAYS TO DEVELOP SYSTEM DESIGNS

Ways to develop systems designs include keeping with the existing system, modifying the existing system, prepackaged designs, and creating new system designs.

Existing System Design

One of the system designs which should almost always be considered is the present system design, that is, to keep the present components, structure, etc. In some cases, minor modifications will have to be made to keep this design viable over the planning horizon.

One of the fundamental things the analyst needs to understand is the key interrelationships within the system under study. Also needed by the analyst is knowledge of how effective the present system is and why there is a problem. Without this information an analyst has little chance of developing and implementing an effective system solution.

Generally, a very useful way of gaining this understanding of basic system interrelationships is by studying the existing system in some depth (see Table 5-1). By studying the existing system an analyst can gain general background knowledge into how the system is presently constructed. The analyst can also gain insights into what the key re-

Table 5-1 Why understand existing system?

1. Insights into system and its key relationships.
2. Gives a picture of where you are.
3. Provides a benchmark.
4. Clues for new alternative solutions.
5. Must know to overcome defender resistance.

lationships are in this area. Very few people actually know how the system really works and how effective it is. By studying the existing system the analyst can see how the system really works, not how people think it works.

The present system also provides a benchmark against which new alternative solutions can be compared. By establishing how much effect the existing solution actually has on various criteria like cost, accuracy, reliability, etc., the analyst can better determine whether in fact new proposed solutions are better and how much so. Because comparison among alternatives is desirable, the existing system can generally be one alternative to be compared against. Clues for new solutions can often be gained by studying how to overcome major weaknesses or disadvantages with the existing system.

The first four reasons in Table 5-1 are well known and are often discussed in texts. The fifth reason, though mentioned rarely, is surely one of the most critical for getting a new system design accepted. The analyst must know the existing system in order to later convince the designers and defenders of the present system. After he has come up with a new alternative solution and it comes time to get top management (CDM) to accept it, the analyst will find he has to sell the proposed system solution. A discussion on how to do this appears in chapter 15, "Communicating the System Study Results." At this point it is important to realize that the analyst's selling effort will be to people who have a reference plane tied to the existing system. They don't know the proposed system, but they know the existing one. So when they ask questions, or make judgments, it is in relationship to what they know (or think they know), which is the existing system. The analyst is doomed to failure if he cannot answer most of their questions or keep their claims (good or bad) for the present system in line with reality.

It is a peculiar thing, but when one is presenting a new idea to a group, the group almost always wants to compare the proposed solution to perfection. Since no solution can stand up to this comparison, the new system looks bad and becomes rejected. The most effective way I know of keeping that from happening is to force the group to compare new solutions to the present solution—not the ideal. While this may sound easy, it is very hard to do and impossible if one does not have a good understanding of how the existing system really works as measured on certain performance measures. Additionally, as will be discussed later, if the analyst cannot answer people's questions he is going to lose the confidence of the group. If their confidence in him is lost, they are not going to buy the proposed system—no matter how good it is!

The understanding of the existing system thus becomes an important part of an effective system study.

Modified Existing System

Another way to develop system designs is to make major modifications to the existing design. Primary emphasis would be on making changes to those aspects of the system which would overcome or at least minimize the negative aspects of the present situation. In this approach, one is keeping the basic framework of the system design and moving in an incremental or evolutionary fashion as opposed to making major fundamental changes which would imply a revolutionary approach. This is generally wise, because as Carl Sagan, in *The Dragons of Eden,* stated concerning the evolution of human systems:*

> It is very difficult to evolve by altering the deep fabric of life; any change there is likely to be lethal. But fundamental change can be accomplished by the addition of new systems on top of old ones . . . Many new organ systems develop by the modification of older systems like the modification of fins to legs, flippers to wings, etc. . . .

In the modification of the existing system approach we are acknowledging the great difficulty of changing the fundamental structure of a system and secondly, the very possible lack of wisdom doing so. With complex systems, it is often not clear what all the consequences will be of making changes to the system. Further, since the system has worked in the past, there is very good reason to try and preserve the system strengths as they have evolved over time. Additionally, in terms of acceptance by the system members of any proposed change, the less the degree of change, the more likely there will be group acceptance. However, conflicting with this thinking is the fact that traditional system designs often can't survive when there are major shifts in the environment. Like dinosaurs, their strengths become limitations under the new conditions. The analyst needs to assess when revolutionary change is required, but generally the evolutionary approach is most effective.

Prepackaged Designs

A different way of developing system designs is to draw alternatives from sources outside your system. The basic thought is to learn from others or other situations and in the process save much time and energy. We will discuss this approach to developing designs considering the methods of "off-the-shelf," "learn from others," and "influential people suggestions."

---

*Carl Sagan, *The Dragons of Eden* (New York: Random House, Inc., 1977, p. 81.

Off-the-shelf: Depending on how common the systems problem is that you are facing, there may well be many predesigned solutions which another company would be happy to sell you. The so-called "off-the-shelf" or turnkey systems are solutions which other organizations have spent much time and money to develop. You most likely can get designs for a fraction of what it would cost you to develop. In reality, this is probably the most typical way of getting a solution. Very few of us design our own alternatives.

In the case of acquiring a place to live, off-the-shelf designs are new tract homes, apartments, mobile homes, etc. If a business had a problem which looked like it needed a computerized solution, there are plenty of salesmen who through personal calls or advertisement will be happy to make you aware of their company's prepackage design, i.e., an IBM computer, a Honeywell minicomputer, etc.

Learn from others: Many new alternative designs can be generated by learning or copying from other people or organizations who are faced with similar problems as you are. In this case, you as a business-person would go and study your competitors to see how they are coping with a particular situation. If there are several competitors, you would try and classify them into two groups: successful and not-so-successful in regards to handling this type of problem. From an analysis of how these two groups differ in their problem-solving strategies, you can determine the key factors in success which you can then incorporate into your own system design.

In learning from others, the least imaginative solution is simply to copy exactly what the other person or organization has done. In the case of computers, if your competitors get a particular computer, then you order the same model. Or if the problem was the more personal one of selecting a college to attend, because your friend picked the local state college, so do you.

Generally speaking, straight copying of solutions from others isn't the most effective. While this method has the advantages of being quick and not requiring any creative thinking, it generally suffers in effectiveness because different systems require different solutions to the same problem.

This is why Leavitt calls this overall learning-from-others approach innovative imitation. To be successful you need to take the solution of others as a starting point, not as the end.

Influential people suggest: As we will see in chapters 13 and 14 there will be a need to consider the views of people who oppose the recommended solution. This is especially true if this person or these people are influential or powerful enough to veto or stall any suggested change. So from a power standpoint, one needs to consider opposing views, but even more often other people who have any understanding of the system will come up with excellent suggestions on various alternative solutions to the problem at hand.

In the case of deciding where to live, when your mother-in-law suggests a house one better listen. You don't have to listen to E. F. Hutton, but your mother-in-law, yes! There are several reasons why the mother-in-law's suggested house should be considered one of alternatives. First, she may have more expertise on living in houses than you do. So you can learn some valuable points. From a power standpoint, to keep future relationships intact, you need to give some reasonable rationale on why you selected another house. This is possible when you show that you considered her suggested alternative, but another house turned out better for reasons a, b, c, etc. Thus, one of the most effective ways of getting a particular systems design accepted by all parties is to evaluate the benefits and costs of a solution in relationship to another solution thought superior by influential people.

In an organizational situation, when selecting a computer to do the basic accounting function it is crucial that you ask the opinion of the office manager. This is especially true if he or she has expertise in the computer field, but also important if the person can effectively block any implementation of the computer into the company.

New Systems Designs

Here we are talking about attempting to determine significantly different designs from those developed through the previously mentioned approaches. While all approaches require creative thinking, the greatest premium would be required here.

Idealized: A potentially very powerful way of generating system solutions which are significantly better is to start by looking for alternative solutions to an idealized system. This approach starts from scratch by assuming there are essentially no restrictions imposed on the system. If the analyst could do anything he wanted, how would he want to design this system? That is, costs are no problem, no external requirements must be met, etc. This type of experience can be very mind-opening. After establishing this framework and thinking through some idealized alternatives, real-world constraints can then start to be added. A rough determination can be made as to how much each constraint lowers the effectiveness of the idealized system solutions. Those constraints which have a major negative impact can then be candidates for extensive probing to see if they are solid or are they just nice-to-have's. The analyst will now have some way to show management the cost of system effectiveness of each of the constraints and the resulting tradeoffs (Ackoff, 1970). These concepts are discussed in depth in chapter 6.

Considering our housing example, an idealized design approach would start out with the concept of designing and building your own custom home. With no restrictions on money, zoning laws, etc., you would work with an architect to sketch a house which would meet all your fantasies. From this base, then external reality would be drawn into the picture and the design would have to be modified to something which would cost less than $200,000 to build, fit on a ½-acre lot, etc.

Parallel situations: This approach involves the concept of cross-fertilization which involves looking for new ideas on how to handle a particular situation in places with no real connection to the system under study except through an analog. For example, many of the more effective ways to function in warfare have come from the study of biology (bionics). The design for the high-speed/low-drag profile of nuclear submarines came from the study of porpoises and their outer skin. Concepts for radar can be understood from studying bats. The study of group behavior in chimpanzees has given some clues on how to structure human task groups. An analyst can gain this type of knowledge or develop these kinds of mental springboards from reading outside his or her field of interest and also from having actual experiences in a wide variety of situations.

Working from the housing analogies of nature, you could come up with highly different home designs. For example, you could consider a beehive-shaped house with a honeycomb structure. You might think of a housing alternative as a basic cave, etc. Remember, with new systems designs we are looking for highly different approaches to help free up our thinking.

Morphological: A third way of developing highly different designs is through an approach called morphology, which is the study of forms and structure. In this approach one needs to define each of the system components (subsystems) and establish the different forms each subsystem could take. Then, the morphological approach to systems design is to determine all possible combinations of subsystem forms. Each combination of basic components is a potential systems design (Advanced Systems Institute, 1976). Because this last approach is much more abstract than the other design methods, let's work through a specific example.

Designing a Chinese Meal: A Morphological Example

Assume you and your family wanted to go out to dinner and try the food at the Chinese restaurant which had just recently opened. After you are seated, the waiter brings you a menu. You are now forced to make a decision on what is the best meal for you. Or, restated in systems terms, what is the best systems design?

For the special dinner at the Jade West, you can pick three entrees (see Fig 5-1). One from column A, one from column B, and one from column C. Column A is the various chow meins to include beef, shrimp, lobster, pork, etc. Column B is the chop suey which includes pork, shrimp, beef, lobster, etc. Column C is the sweet and sours category which includes pork, beef, shrimp, etc.

If one used the morphological approach to this question of designing a Chinese meal, he would want to first establish what are the components of the meal. In this case it would be the various types of chow mein, chop suey, and sweet and sours. Then considering the various forms or types each component could take, one would consider all the possible combinations.

Each of the combinations would be a particular design of a Chinese meal. For example, one design would be beef chow mein, shrimp chop suey, and sweet and sour pork. A second design would be beef chow mein, shrimp chop suey, and sweet and sour shrimp. A third alternative would be beef chow mein, beef chop suey, and sweet and sour beef.

From this simplified example, one can get the main thrust of morphological designs of systems. However, hopefully one can also see its main limitation. How many possible Chinese meals are there? Except in extremely important questions, one wouldn't want to go through this complete brute force approach to designing systems. However, the concept is excellent for getting analysts to free up their thinking and

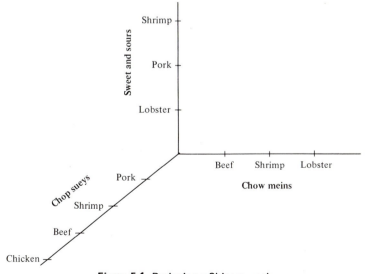

Figure 5-1 Designing a Chinese meal.

develop several ideal designs they most likely wouldn't have considered otherwise.

Because this morphological method can be a very useful tool in a creative sense, let's develop a simplified notation for this process borrowed from mathematics. In the case of the Chinese meal, it was a function of three major components or variables, i.e., column A—chow meins, column B—chop sueys, and column C—sweet and sours. Therefore:

> Chinese meal is a function of the combinations of (column A, column B, column C)

> or

> Chinese meal = f (chow meins, chop suey, sweet and sours)

> or

> System solution = $f(x_1, x_2, x_3)$

This is where x_1 is a variable which stands for all the different chow meins, etc. If there were 25 different chow meins, 20 chop sueys, and 20 sweet and sours, how many Chinese meal designs would be possible?

$$25 \times 20 \times 20 = 10,000$$

Let's use this morphological notation to design a solution for determining what is the best car for the Jones family. What are the *major* components of a car?

$$\text{Car} = f(x_1, x_2, x_3, \ldots x_n)$$

where x_1 could be   Make
        x_2 could be   Model
        x_3 could be   Year
        x_4 could be   Engine
        x_5 could be   Transmission
        x_6 could be   Color
        etc.

Therefore, a car design is:

> Car = f (make, model, year, engine, transmission, color)

Some potential car designs would be:

Car_7 = Toyota, Corolla-2 dr., 1978, 1600 cc, automatic, blue
Car_{212} = Ford, Pinto station wagon, 1982, 1200 cc, 4-speed, brown

Hopefully, you now can start to see the power of this technique. There are literally millions of car designs available to a person today. A far cry from the early 1920s when Henry Ford gave a choice of one car, i.e., Ford, Model T, 1920, 4 cylinders, 3-speed shift, black. With increased options comes the need to make choices. Life was simpler with just one basic solution, but not nearly as effective and efficient as being able to tailor a solution to your particular system needs.

Thus a systems design is really related to all the components shown in the systems context diagram. In the case of the Jones family considering getting a car (Fig. 2-1), to reduce the complexity of the problem most of the components are held constant.

Cascading: In the discussion of morphology above, it became quite clear that while this method is excellent for developing some creative ideas, there are just too many potential solutions to really consider—even in the simplest of systems (e.g., Chinese meal).

An approach which takes advantage of the morphological process, yet reduces the possibilities to a reasonable scale is what we call the cascading approach. As an example, we return to the problem of selecting a family car. The basic notation is:

Car = f (make, model, year, engine, transmission, color)

We know the possible combinations are in the millions. Since we can't examine them all, in the cascading approach we arrange the components (variables) in what we consider the order of importance. In the case for the Jones family, let's say after some study it worked out as follows:

Car = f(year, model, color, transmission, make, engine)

We take the first variable the year as most important, then the model, etc., down to the type of engine.

The variable year can take on values such as 1982, 1981, 1980, . . . , 1922, etc. All the years that cars might be available today. If after studying the Jones family we conclude that they should limit their choice of cars to new ones (i.e., 1982), then we can eliminate all other year possibilities.

Then in the cascading approach we go to the next variable model considering only new 1982 cars. Considering the number of kids the Joneses have, the pet dog who they like to take for drives, etc., they

Table 5-2 Summary of system design methods.

Method	Housing Example
Existing system	Present house—three bedroom
	New rug
	New dishwasher
Modified existing system	Remodel present house
	Add-on fourth bedroom
	Repaint outside
Prepackaged design	
Off-the-shelf	New housing tract
Learn from others	Buy neighbor's four-bedroom house
Influential people suggest	House mother-in-law suggests
New systems designs	
Idealized	Design customized house
Parallel situation	Build beehive house
Morphological	Consider house, apts., condo., mobile home
Cascading	Select home by bedrooms first, then family room, then backyard, etc.

feel they must have either a station wagon or a four-door car. All other models are eliminated.

For color suppose the Joneses say they would like a light blue, light brown, or a silver gray. Other colors are not acceptable. Because of the gasoline situation, Mr. and Mrs. Jones feel a four- or five-speed shift transmission is needed.

The make of the car has to be American and have a service dealer within 20 miles of the Joneses' home or where each works. The engine needs to have a high enough power-to-weight ratio to provide safety on entering freeways. This has been established as at least a four-cylinder engine with 2000 cubic centimeters displacement.

Given this background on the Jones family, let's see what we have:

Jones Family Car = f(1982; station wagon or four-door; light blue, light brown, or silver gray; four- or five-speed shift; American made and local service; at least four cylinder with 2000 cc).

By studying the Joneses' family situation, we have been able to reduce the possible alternatives (as generated by a pure morphological approach) from millions down to hundreds. In turn, we can reduce this even more if we desire.

When we go to look for specific alternatives, we only need to consider new 1982 cars. Of all the new cars we need only consider those that have station wagons or four-door models, etc. We cascade down from the most important variable. Note that we did not try to get just one entry for each variable. That is, we didn't select the best model, best color, best transmission, etc.

WHY ALTERNATIVE SOLUTIONS ARE NECESSARY

A major point in the philosophy of the systems approach is that alternative ways of solving a problem must be considered before one can make a good decision on which solution to implement. Why is this so?

Conditions of Uncertainty

The necessity for alternative solutions is directly related to how much can be known about a problem. If in fact the situation is one where there is "a solution" in the traditional sense of that phrase then it is a waste of time to be considering alternative approaches. One should search until he finds "the" solution.

However, as we noted in chapter 4 on the complexity of systems problems, the set of problems which have "a solution" is very small. Further, the approach of this text is designed to help guide the analyst in tackling much more complex systems problems. Additionally, the types of problems we are considering require thinking about the future (i.e., planning horizon). This introduces considerable uncertainty into the decision process, since there are no facts in the future, but rather just assumptions, conjectures, and guesses.

Because of the limited resources of time, manpower, etc., all facets of a system can't be checked out. Additionally, the main players (i.e., analysts, managers, workers, etc.) are all humans who have varying perspectives, biases, expertise, etc.

We see therefore that, given the types of problems we are considering (under conditions of uncertainty), one could easily be wrong in the solution that is selected. Often, in fact, we won't know until some time after the solution is implemented whether it was really a good decision! In most cases, this is unfortunately too late. Better answers need to be found prior to reaching a decision.

One of the main ways of increasing your chances of picking a good system solution, under conditions of uncertainty, is to develop and compare alternative ways of solving a problem.

Relative vs. Absolute Performance

One of the major points that we make in this book is that systems problems are not solved in the dichotomous sense of correct/wrong. Rather the "solving" of a systems problem is related to whether the problem is basically operational, tactical, or strategic. If it is operational, then problem-solving serves to make the system viable. Tactical problems require solutions which increase the efficiency of a system, whereas solutions to strategic problems relate to increasing the system's effectiveness (recall the discussion in chapter 4, Table 4-5).

Efficiency and effectiveness, we will see in chapter 8, are really concerned with systems performance which in turn will be measured by values, objectives, criteria, etc. But values, objectives, and criteria are continuous concepts. There is no end to the desirability of wanting better performance and to minimizing negative aspects of systems. In other words, how much is enough? How much money would you like to make? No matter how much money you are presently making, you always want 10% more. It seems a bit like the greyhound dogs at the race track chasing the mechanical rabbit. However fast the dogs go, the rabbit always goes just a little faster.

When dealing with continuous things which also have positive and negative values attached, there are no absolutes. One is forced to use a comparison base. In other words, performance is not good or bad per se, but can be only measured in comparison to some standard. The standard which we will use is to consider what else is available. That is, what level of performance is generally feasible in a particular situation?

The specifics on how to deal with uncertainty and how to establish relative performance will be dealt with in chapters 8 and 12. For our purposes at this stage, we only need to recognize that in dealing with systems problems we have to replace the mathematical concept of one correct solution to a problem, with the notion of many solutions . . . some which are better than others.

THE NUMBER OF ALTERNATIVES TO CONSIDER

We have stated that alternative solutions must be considered. But we have not answered the question of how many alternatives need to be compared. The number of alternatives that should be considered before making a decision as to the best alternative is a function of which basic philosophy of problem-solving you follow.

The two most prevalent philosophies in the literature are *satisficing* and *optimizing.*

Optimizing is a method of problem solving in which *all* viable (feasible) alternatives are compared. The best of these is accepted as the desired solution. Consequently, there is no better system solution.

The literature of operations research, management science, and quantitative analysis in general stresses that only the optimal solution to a problem should be accepted as the best solution (Dantzig). Following this philosophy one should only make a decision after he has considered all possible workable alternatives.

Satisficing is a method of problem solving in which the *first* alternative that meets all restrictions (constraints) is accepted as the desired solution.

Cyert and March in their study of the behavior within organizations, showed that decision-makers rarely if ever optimize but most generally satisfice. That is, most decision-makers accept the first workable solution to a problem.

Which philosophy is correct? If one makes decisions by satisficing, he need only consider any one alternative that is feasible. Whereas, with optimizing all feasible alternatives must be considered (which could be in the millions!). The answer to which is correct is that neither is. Instead, we suggest a third philosophy of problem-solving—*adaptavising* (Ackoff).

Adaptavising is a method of problem-solving in which one continues to check viable (feasible) alternatives until the perceived costs of further search equal or exceed the potential benefits to be gained from a better system solution.

The concept of adaptavising is offered here as both a preferred philosophy of problem-solving and as a means to better explain what decision-makers should do in practice.

Adaptavising implies that one is not committed to the idea of checking out all alternatives or selecting just the first feasible solution. The analyst who adaptavises adjusts his search for the best system solution depending on how important the problem is and what hope he has of determining better system solutions.

As a general strategy, one who follows the adaptavising approach would check out several feasible alternatives. After an evaluation has been made and the best of these alternatives selected, a judgment is made as to the potential increased benefits possible from determining an even better system solution. The amount of increased benefits is then compared with the costs of further search. These costs are to include not only dollars, but also time, energy, and frustration. If the perceived benefits from further search exceed the costs, then more alternatives are determined and evaluated. If the costs exceed or equal the potential benefits from further search, the best alternative so far is then selected as the preferred solution.

As a practical rule, one must always have at least two solutions, one of which should be the existing system design. A problem has to be very important before it makes sense to evaluate in depth more than five alternatives.

In this text, we will adopt the adaptavising strategy as the primary basis for determining the numbers of alternatives to consider before making a decision as to which system solution to implement.

ENHANCING SOLUTION DEVELOPMENT

The developing of alternative system designs is a highly creative activity. The thought process required to "invent" solutions to a problem is very different from the thinking necessary to establish what the problem really is.

To aid in the invention of alternative system designs, we have discussed the specific techniques of considering the existing system design, modifying the existing system design, utilizing prepackaged designs, and developing conceptually new designs using the idealized, morphological, and cascading methods. These techniques will no doubt prove helpful in this regard. However, one must also have some understanding of the creative process and what conditions either aid or disrupt the process.

Osborn in his book, *Applied Imagination*, suggests that for effective brainstorming a group of about twelve people is ideal. This group should be given a very clear and concise task statement of what problem they are seeking solutions to. Several relatively brief periods of 30 to 45 minutes should be allocated for solution development and deadlines or quotas should be set. In other words, fifty solutions in thirty minutes. After several days have gone by, another 30-45 minute brainstorming session should be scheduled. If possible, some group members should be replaced to make sure that new perspectives are inputted to the process and the group doesn't get hung up on stale combinations. Also the setting should be some type of "retreat," as opposed to the usual work offices, since a different surrounding fosters more nontraditional thinking.

We have listed some of the conditions which have been shown to make for an ideal brainstorming setting. For idea generation to really be effective, (1) criticism is ruled out; (2) free-wheeling is welcomed; (3) quantity is wanted; and (4) no pride of ownership.

Table 5-3 Summary of problem-solving philosophies.

Literature Name	Number of Alternatives	Concept
Satisficing	1st feasible solution	Good enough
Optimizing	All feasible solutions	None better
Adaptavising	Reasonable number of feasible solutions	Best of available

The quickest way to bring the flow of ideas to a halt is for the group to criticize a person for his thoughts. All judgment as to the worthiness of an idea must be deferred to a later stage. Free-wheeling or the saying of whatever comes to mind must be encouraged. The wilder the better, since this encourages others to open up their minds to highly different approaches.

The greater the number of ideas the better, since quantity tends to breed quality. Somebody needs to write down the ideas as they are generated or they can be taped. When the mind sees that ideas are recorded, it is then freer to go on to other thoughts. In general, pride of ownership of the ideas needs to be minimized so others can take an initial thought and improve on it by combining it with other ideas, etc.

Basically, in this whole brainstorming process you are trying to set up a chain reaction of thoughts among group members to derive better solutions than could be done by them individually. Additionally, for people to be at their creative best, there generally has to be pressure on them to come up with solutions (i.e., necessity is the mother of invention).

It is very important to appreciate that brainstorming is concerned with ways of generating potential solutions, not with what we will call feasible or workable solutions. A weakness of the general creative process is that people then don't take these "wild" ideas and apply the hard judgment necessary to test their worthiness. This we will discuss in the next chapter.

Class Exercise 5-1

HANDLING INCREASED STUDENT DEMAND

The school board for the San Juan Capistrano schools has tried to finance the construction of additional "conventional" school buildings to meet the ever increasing number of students moving to this fast growing area. However, the last two attempts for bond financing have been defeated by the electorate.

Because students must be given an education, the board of trustees studied alternatives for increased capacity ranging from pitching tents to negotiating the use of a virtually vacant Federal government building located nearby.

Alternatives presented to the trustees by the superintendent were: (1) air rights (the ownership or sale of space above land or existing buildings); (2) air structures (a plastic bubble supported by slightly compressed air); (3) conversion of existing spaces (use auditoriums, etc., as classrooms); (4) mobile facilities (such as a library in a van, which would free up existing rooms for classes); (5) extended school day; (6) extended school year; (7) double sessions; (8) aid from developers, both voluntary and mandatory; (9) residential schools (classes held in unfinished new housing); (10) Federal Building in Laguna Niguel; (11) tents; (12) outdoor teaching spaces; (13) the Regional Occupational Program (ROP) building modular classrooms; (14) a successful bond election; and (15) a successful lease-purchase election.

1. (a) What is the system?
   (b) Who are the CDM's?
   (c) What is the problem?
   (d) What is the planning horizon?
   (e) What is the cause of the problem?

2. A brainstorming session produced the above-mentioned number of alternative solutions. Notice the wide range of scope. Group the alternatives into the classifications of
   (a) existing system
   (b) modified existing system
   (c) prepackaged designs
   (d) new system designs.

3. (a) Having seen some of the alternatives, now what do you think the problem is?
   (b) If this is different from your answer in question 1(c), Why?
   (c) If not, why not?

4. Put the alternatives in the Chinese menu notation. What are the basic components? Which alternatives, as described in the newspaper quote, are not complete?

Class Exercise 5-2

SMALL BUSINESS COMPUTER CONSIDERATIONS

In the explosively growing mini-micro computer world for use in small businesses, there is a very wide range of options available. From the businessman's point of view this hodge-podge of computers and services is confusing. But in a Chinese menu sense, all alternatives boil down into the three components of (a) computer, (b) applications software, and (c) maintenance service.

For the following options, first put them into Chinese menu notation and then further classify them by existing, modified existing, prepackaged, or new system designs.

1. A microcomputer manufacturer sells personal computers through retail electronics stores. Very limited software is available. The customer is expected to write their own software programs. For maintenance, the businessman must bring the computer into the retail store.

2. A major computer manufacturer sells microcomputers through company salespersons. Standard application packages are sold with no modifications allowed. Nationwide service is available at the location of the business.

3. A major minicomputer manufacturer sells business computer hardware. No software or maintenance is available.

4. A turnkey software house gives you applications packages tailored to the specific needs of a business, provides the appropriate computer hardware, and services the hardware and software maintenance needs.

5. A business can buy hardware from the manufacturer. He then is given a list of software houses to contact for appropriate application program needs. Alternatively he can contact an independent hardware maintenance company.

Class Exercise 5-3

STARTING A NEW BUSINESS

Many an employee has aspirations of going into business for himself. Discuss the following options in terms of their inherent advantages and disadvantages. Then classify them by existing, modified existing, prepackaged, or new systems design.

The aspiring businessperson is considering:

1. Opening a new fast food place featuring a new food recipe that he has invented.

2. Buying out a successful restaurant that has been in the same location for 20 years.

3. Opening a new restaurant that is part of a nationwide franchising chain.

4. Taking over the retail mini-market that he has been working at, and adding take-out luncheon sandwiches for the nearby office workers.

5. Increasing the operations hours of a fast food operation he owns to now include breakfast, in addition to the regular lunch and dinner meals.

Chapter Six

Establishing Which Alternatives to Evaluate

As we have seen from the discussion in the previous chapter, a wealth of alternatives can usually be derived for any systems problem. However, as we noted in the section on brainstorming and also in the class exercise in chapter 5 on handling increased student demand, the premium was first placed on generating ideas. How good or workable the ideas really are was not addressed. Additionally, we stated that the philosophy we were going to follow in determining the number of alternatives to evaluate, was the adaptavizing approach versus either satisficing or optimizing.

So at this point in the systems approach to problem-solving, we have two basic concerns. First we have generated a "wild" collection of alternatives, but without determining which are really workable solutions to the systems problem at hand. Secondly, after we have weeded out those nonworkable alternatives, we may still have too many alternatives to evaluate. How does one get down to a manageable number of options?

In this chapter we will address these issues by first discussing the concept of constraints which can be used to eliminate nonworkable alternatives. Then we will cover how to determine those potential alternatives with the highest probability of success.

CONSTRAINTS

The determination of what constraints are or will be placed on a system is a necessary step in good system design. In this respect, it is helpful to view constraints along the dimensions of time, control, solidness, application, and types.

> **Constraints** are restrictions and/or requirements placed on a system which *must* be met in order for the system to be viable.

Time

Constraints can be looked at as those requirements which are presently demanded of the system and those which will be required some time in the future. For the purposes of a system study, the future is defined to be the assumed planning horizon under consideration. As an example of the time dimensions of constraints, assume you were asked by Yellow Cab Company to determine what kind of cars they should buy for their Los Angeles operations for the 1982-87 time frame. Among your considerations would be the changing emission standards for cars used in the Los Angeles area as required by law (Clear Air Act). Most American car makes and models (system solutions) which would meet the 1981 emission standards could be unacceptable (infeasible) under the 1985 emission guidelines. Therefore, system designs must consider constraints placed upon the system at any time during the total planning horizon.

Control

Another important consideration of constraints is the determination as to whether or not the CDM has or could get control over these restrictions. If the requirements are outside of the control of the CDM, then they are considered external constraints and in most cases have to be taken as a given. If the restrictions have been imposed by the CDM or are under his area of authority they are called internal constraints. There is generally a much greater chance of getting these internal constraints changed, especially if it can be shown a better system solution will result.

> **Internal constraints** are restrictions placed on the system which must be met, but are within the authority of the CDM to change.

External constraints are restrictions placed on the system which must be met, but are not within the authority of the CDM to change.

Using the Yellow Cab example, the emission standards for autos required by state law in 1985 is an example of external constraints. Unless Yellow Cab plus others have tremendous political muscle, their taxi cabs will have to meet this standard in 1985. On the other hand, assume the president of Yellow Cab wants to show his company is socially responsible. The president feels that Yellow Cab should help clean up the smog condition in Los Angeles now and not wait to 1985. Therefore, he is requiring that all Yellow Cabs must meet the more stringent 1985 emission standards in 1981. This requirement is clearly an internal constraint under the discretion of the CDM. After viewing several alternative solutions, assume a systems analyst determines that requiring all Yellow Cabs to meet the 1985 emission standards by 1981 will result in excessive costs. These additional costs will lead to a dismal corporate profit picture for the company in 1982 and 1983. What will happen to the more stringent requirement? The president will probably rescind it, whereas the external constraint of the 1985 emission standards must be met in 1985 through 1987 whether it results in dismal profits for Yellow Cab or not.

Solidness

A third factor to consider in designing system solutions is the determination of whether the constraints that have been imposed either internally or externally to the system are in fact solid or just desirable. In this case solid means the constraints must be met; desirable refers to constraints are merely nice to have. In many systems of a technical nature, there are natural or physical laws which impose certain restrictions. For example, in the designing of airplanes or their airborne weapon systems, the frictional forces of drag must be contended with as a natural phenomenon not a whim.

Restrictions placed on a system which are of a legal nature many times must be considered as solid. For example, assume you were designing an inventory system for a furniture manufacturer. The law requiring the firm to pay taxes on the value of the inventory in stock on March 1 of each year is a given. The taxes must be paid according to the schedule stated in the law. Because the payment of taxes is a law *and is enforced*, any inventory system design must consider this tax requirement as a solid constraint (i.e., it must be met as stated).

On the other hand, much of what are considered as system requirements are in fact nice to have or in many cases just whims of management (or others). Thus, before the analyst should accept restrictions placed on his potential system design, he should test the strength of conviction of these management-imposed requirements to determine which are solid requirements and which are just desirable. Many of these system requirements have come about in the present system because somewhere some analyst asked people what they would like to see in a system.

In designing many management information systems, analysts go to each manager and ask what kind of information he needs or would like to see in various reports he gets or should be getting. What in fact they get is a dream sheet of information requirements. As this dream sheet is compiled and works its way through the various levels of the chain of command, it comes out as solid system requirements that must be met for the information system to be feasible.

Why isn't this procedure reasonable? First of all, people generally do not know what kind of information they need to do their job without some real good analysis. Secondly, if there is no cost or penalty attached to excessive information demands, people will have a tendency to ask for everything to be on the safe side. Since it does not cost anything, why not? Thirdly, information requirements are not a one-time determination. The information requirements change as the job content changes, as new knowledge becomes available, and as different people rotate into a job holding much different perspectives. The overall point of this example is to show that requirements are not necessarily what people say they are; that one needs to analyze the process to really determine what must be produced.

Application

Most often when constraints are defined, they apply to the system and therefore affect all potential alternatives. That is, for any alternative to be considered a feasible solution for a particular system, it must meet all constraints. Thus, the constraints are considered as general. In other situations, the constraints will only apply to a certain class of alternatives. In these cases, the constraints will be considered as specific.

For example, in the determination of your eligibility for attaining financing for the purchase of a place to live, you will be required to make a down payment of 10 percent for a housing loan. Whereas the same bank will require you to make a down payment of 20 percent for a loan on a mobile home. Thus, these external constraints are specific to the class of alternatives.

An example of a general constraint was in the late 1960s when the airlines decided to install a computerized reservation system. They placed the requirement that any computer system to be feasible must give a response time of less than three seconds.

Types

It is also helpful in determining the requirements of a system to think in terms of the categories of technical, economic, psychological and political. This classification scheme is covered in depth in chapter 7. So at this point we will just note the four major types.

A summary chart of system constraints appears in Table 6-1.

Table 6-1 Summary chart of constraint classification.

Control	Solidness	Time	Application	Type
External	Enforced	Now	General	Technical
				Economical
Internal	Nice to have	Later	Specific	Psychological
				Political

POTENTIAL VS. FEASIBLE SOLUTION

To aid in our need to screen system designs that have been offered as solutions, we need to define the difference between potential and feasible solutions.

Potential solution is a systems design which appears likely to solve the stated problem.

But potential solutions are not necessarily solutions which are workable. Study is necessary to identify the feasible solutions.

Feasible solution is a systems design which will solve the stated problem and will meet all constraints (restrictions).

What we are really interested in evaluating in depth are feasible solutions.

The first screening test is to see which potential solutions meet all applicable restrictions that have been specified. For those designs which don't, they must either be eliminated or modified so that they will meet the constraints and thus become feasible or workable designs.

Determining Feasible Alternatives Example

Let's work through a full example to show how the various ways of designing and screening alternatives might work. In this case, let's assume the problem is "What is the best graduate school for Jim Brown?"

With the problem stated in this fashion, note the implications for the development of various solutions. We would tend to look at the various colleges with graduate programs. What if the problem had been stated to be "What is the best way for Jim Brown to get an advanced education?" Now the alternatives could be, in addition to the various colleges with graduate programs, a plan for educating oneself at home through a disciplined reading program, immersing oneself in a wider variety of assignments at work, attending various one-week seminars on advanced topics, etc.

Further, note the difference of focus if the problem had been stated as "What is the best career for Jim Brown?" Here the alternatives could include: (1) continuing to do what he is doing (i.e., working for the same company in the same career area); (2) change career paths but stay with the same company; (3) quit work, prepare for a new career field by attending graduate schools, etc. These examples of problem statements are used to once again emphasize the relationship between problem identification and the development of alternatives.

For the rest of the analysis, we will stick with the problem statement, "What is the best graduate school for Jim Brown?" In remembering the discussion on developing alternatives and brainstorming from chapter 5, we decide this problem is very important to the future of Jim Brown so we are going to develop the widest selection of schools to pick from.

Thus, we go to the library and check a reference book like Lovejoy's *Guide to Graduate Schools.* The first thing we notice is that there are over 100 pages on colleges with graduate programs, certainly far too many to really evaluate. How can we screen out those graduate schools which aren't viable for Jim Brown? That is, all the graduate schools mentioned are prepackaged designs which are potential solutions to Jim Brown's problem. But we need to determine the difference between potential solutions and feasible solutions.

Therefore, we need to set up some constraints. The constraints are derived from the systems context which in this example is Jim Brown, or if he has a family, it would be the Jim Brown family. In talking with Jim and other members of the family, we note that he doesn't just want to go to graduate school but rather has already determined (assumed) he wants to get a Master's in business administration to go with his undergraduate degree in engineering.

So the first constraint would be that the college must have a graduate program in business offering an MBA. This constraint would eliminate a large number of colleges. Note, this could also have been handled through tightening up the problem statement to read "What is the best MBA program for Jim Brown?"

Let's assume the new problem statement is now "What is the best MBA program for the Jim Brown family?" By getting a clearer focus on what the problem really is, many potential solutions will be eliminated from further consideration. In this example, we can exclude all those graduate programs in engineering, arts, etc.

At this point, assume we still have a very extensive list of potential solutions to solve Jim Brown's problem. These alternatives are represented by the following list (shortened for convenience), San Diego State, Arizona State, Harvard, Stanford, Texas Tech, Florida State, Puget Sound, La Verne, USC, San Francisco State, Cal Poly, Claremont, Illinois, Alabama, etc.

We need to analyze the admission requirements of the various schools. These are usually stated in terms of grade point average (GPA) minimum and a minimum score on the Graduate Management Aptitude Test (GMAT). Assume Jim had a 3.05 undergraduate GPA and a 500 GMAT score. Thus, we need to check which colleges Jim's GPA and GMAT scores exceed the specific external constraints.

These constraints will serve as a filtering device (see Fig. 6-1). All the potential alternatives are loaded into the hopper. The filters have "holes" which allow the still feasible alternatives to pass through. The alternatives that do not meet the constraints are discarded as not truly workable solutions.

Assume after checking that the alternatives that met the first test were San Diego State, Arizona State, Texas Tech, Florida State, Puget Sound, La Verne, USC, San Francisco State, and Cal Poly.

Next, the Jim Brown family financial situation is looked at and his potentiality for financial aid through loans and working is considered. It is decided that the maximum tuition that Jim Brown can afford is $1000 per year. This internal constraint is then applied to the remaining potential alternatives. This filtering restriction removes the private schools of Puget Sound, La Verne, and USC. Note that while Alabama is less expensive than $1000 per year in tuition, it is not considered since that alternative was dropped from further consideration after the admission constraint. That is, to be a feasible solution, an alternative must meet *all* internal and external constraints. If at any point it can be shown that an alternative doesn't meet a constraint, it is removed from the list. It isn't worth checking out how it does on other constraints.

At this point, our set of potential alternatives which have met the eligibility constraint and the tuition cost constraint are San Diego State, Arizona State, Texas Tech, Florida State, San Francisco State and Cal Poly. Jim Brown at this point sits down with his future bride and they discuss any restrictions that should be placed on the graduate school decision. Since they will be getting married in three months, the CDM's are Jim and his future wife. They decide because of their love for the West and the need to be near his very sick mother, that only schools

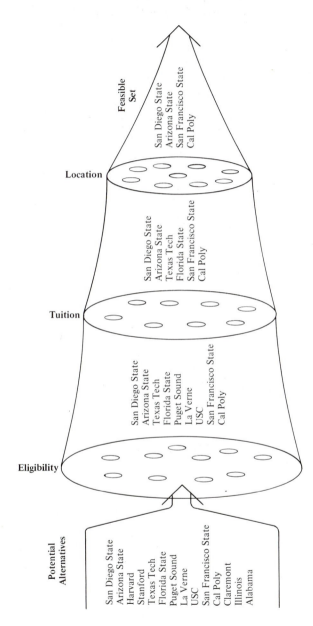

Figure 6-1 Screening for feasible alternatives.

west of the Rockies should be considered. Applying this internal constraint removes Texas Tech and Florida State.

For our example, assume that these are all the constraints that apply. The feasible set of alternatives is, therefore, San Diego State, Arizona State, San Francisco State, and Cal Poly.

Let's review what we have done in this example. We first tried to get as many potential solutions to the problem as possible. However, the number of alternatives was overwhelming. So we went through a screening of the alternatives to determine which are really feasible solutions as opposed to just potential solutions. We applied an eligibility constraint of GPA and GMAT which is external and very likely can't be changed. Next we applied an internal constraint from the systems context of the Jim Brown family of tuition and then location. Thus, we were able to consider literally hundreds of potential solutions, yet got down to a workable set of four alternatives in this case.

Note in this example the constraints were stated very specifically.

Eligibility: College *must* accept student with GPA of 3.05 and GMAT of 500.

Tuition: College tuition for the year *must* not exceed $1,000.

Location: College *must* be located west of the Rockies.

Words like prefer, wish, would be nice, etc., are not used in specifying a constraint. Must implies a yes or no answer. The solution either meets the constraint or it doesn't. If it does okay; if it doesn't, the alternative is either dropped or modified until it meets the restrictions.

SCREENING ALTERNATIVE DESIGNS

The first major test in determining which alternatives to evaluate in depth is to establish which are feasible. These concepts and considerations have been discussed in this chapter under the heading of constraints. From the feasible alternative set we then can apply the screening techniques of major criteria and multistage sampling.

Major Criteria

If after identifying which alternatives are feasible, the number of designs remaining still needs to be reduced, one can attempt to predetermine which of the feasible solutions will have the highest overall worth.

The screening at this point is different from determining potential vs. feasible solutions. All solid internal and external constraints which would apply over the planning horizon must be applied to potential solutions to determine whether in fact these alternatives are workable. Therefore, all feasible solutions meet the constraints and thus the constraints are of no more value in selecting among the alternatives.

In determining the relative worth of feasible alternatives, one needs to use system objectives or criteria as will be described in detail in chapter 7. At this point, objectives or criteria will be considered values important to the system in regard to the alternatives. In the graduate school example, criteria could be things like quality of program, job placement potential, time required to complete the program, etc.

Generally speaking, for an alternative to do well in the final overall evaluation (chapter 12), it has to score fairly high on the most important criteria. Thus, a means of eliminating some of the feasible alternatives in the early stages of analysis is to see how well these alternatives perform on the two or three most important criteria. Even though criteria are continuous, we can artificially set up lower limits on their acceptability. With this scheme, the criteria become, in effect, constraints which can be used as a screening mechanism.

After you have learned the concept of Systems Utility Function as it is discussed in chapter 12, you will see that we can set these lower limits of performance at the barely acceptable, below average, or even the average level of performance.

In using the concept of major criteria screening, it is very important for the analyst to recognize that the artificial limits imposed act like internal constraints. These constraints can be changed or removed as desired or needed. Additionally, the higher the level of performance set, the more powerful the screening device; that is, the smaller the number of *really* feasible alternatives getting through the sieve.

Remember, we are working to increase the probability that we will evaluate a limited set of feasible alternatives which have the greatest chance of being most effective system solutions. But we could be wrong by eliminating perfectly good solutions! We knowingly take this risk because of the limitation of time, money, and manpower needed to evaluate all feasible alternatives (i.e., where that number is very large).

MultiStage Sampling Approach

If after applying the screening techniques of constraints and major criteria, there are still too many alternatives to reasonably evaluate in depth, one can go to the concept of multistage sampling. The first is to classify all the remaining feasible alternatives into major categories. These categories may be natural groupings or you might have to establish a scheme yourself.

For example, in the problem of selecting a car, the remaining cars could be Honda Civic, Toyota Corolla, Toyota Celica, Chevy Vega, VW Rabbit, VW Bus, Dodge Van, Pontiac Firebird, etc. The first level grouping could be:

Group 1	Honda Civic, Toyota Corolla, VW Rabbit
Group 2	Chevy Vega
Group 3	Toyota Celica, Pontiac Firebird
Group 4	Dodge Van, VW Bus

In this case the grouping might have been done on price or it could have been done on function. Once the first level classification is performed, then you need to select *one alternative from each grouping.* Each alternative selected will be looked at as a representative sample of what the performance and cost will be for that category. In selecting which alternatives to pick for each group, the analyst should try to select that alternative which appears like it might be the best overall. Note, that you can't know this for sure until after you have completed a study. But the purpose of this section is to help you when you can't evaluate all the alternatives.

Continuing the example of car selection, assume the first stage sampling yielded:

Group 1	VW Rabbit
Group 2	Chevy Vega
Group 3	Toyota Celica
Group 4	Dodge Van

What you are in fact saying with this selection is when you speak of economy subcompact cars, you are going to use the VW Rabbit as a very good choice on showing the relative performance and cost. Further, to see the performance and cost of vans, you are going to look specifically at the Dodge Van.

The second stage is to do a relatively quick evaluation of the representative sample alternative for each category in terms of criteria that have been developed. The purpose of the evaluation is to see the relative desirability of each category of alternative.

The third stage is the final selection of the alternatives to evaluate in depth. This in turn depends on the outcome of stage two. If the stage two evaluation resulted in any of the categories performing very poorly, these groups should be eliminated from consideration in stage three. Conversely, if one group performs exceptionally well by dominating all other categories, then the final alternatives should be selected strictly

Table 6-2 MultiStage sampling approach.

Stage 1	Stage 2	Stage 3
Initial grouping of alternatives	Quick evaluation on major criteria	Selecting alternatives to evaluate in depth
Selecting representative sample		

from that particular group. A third outcome could be where all the alternative groups perform equally well. In this case, all groups should be represented in the final stage three evaluation.

Continuing with the car example, assume in the stage two evaluation it was shown that group 4 dominated all other categories. Then the purpose of the study could be refined to "What is the best van for the Jones family." The alternatives would be limited to vans and the original list of vans expanded as appropriate.

On the other hand, assume the groups, as represented by a sample, perform equally well in the quick evaluation against the criteria. Then the stage three alternatives should contain an alternative from each category. Further, the specific group alternatives could be the same as in stage two or individual alternatives could be changed. In the process of performing a study, new information is gathered and the analyst and managers become more knowledgeable. Thus, different alternatives can look more or less attractive as time goes on.

A summary of this approach to developing a limited set of feasible alternatives is shown in Table 6-2.

SUMMARY: WHICH ALTERNATIVES TO EVALUATE IN DEPTH

In chapter 5, we discussed why alternatives are necessary and how the number to be evaluated depends on one's particular problem-solving philosophy. A case was made for using the adaptavising approach to trying to determine the "best" of a reasonable number of solutions.

The actual number of alternatives to be evaluated in depth depends on the importance of the problem, how much time is available for studying, and the number of feasible solutions there are. Generally speaking, the number of alternatives to be evaluated in depth should be at least two and no more than five.

In being consistent with the adaptavising philosophy, while it is recognized that all alternatives can't be checked out, there is the desire to try and cover as wide a range of solutions as is practical. To accomplish this task, it was suggested one should be very creative in the initial stages and try to develop both a large number of alternatives and ones which are highly different. Methods suggested were covered under the topic headings of existing system design, modified existing system design, prepackaged designs, and new designs.

In chapter 6, we have considered how to go from an overwhelming number of "wild" potential solutions, to a limited set of alternatives which have the greatest chance of proving to be excellent answers to our systems problems. One screening method is to use constraints to eliminate potential solutions which aren't feasible. Secondly, if the number of feasible alternatives is too great, they could be reduced by using major criteria as screening devices. Lastly, the concept of multistage sampling was suggested as an especially good approach for zooming in on the best of a large number of highly different solutions.

By way of a summary to chapters 5 and 6, the general points listed in Table 6-3 can be applied to establishing which alternatives should be evaluated in depth.

These thoughts are given with the understanding that while it can't be predetermined for all cases which types of alternatives should be evaluated, some general principles are helpful in guiding the analyst in this crucial phase of problem-solving. The more important the problem to the system, the more alternatives one should try to evaluate in depth. But this is always at the cost of more time, energy, and dollars.

Table 6-3 Deciding which alternatives to evaluate in depth.

- Screen alternatives for feasibility.
- Existing system design should be considered whenever possible.
- Consider design suggested by influential people.
- Have at least one highly unusual approach.
- Select an alternative from each of the most promising groups.
- Limit overall number of feasible alternatives to between two and five.

Class Exercise 6-1

SELECTING A GRADUATE SCHOOL

Using the example of Jim Brown deciding which graduate programs to evaluate (Fig. 6-1), answer the following questions:

1. What would you do if the final set contained 40 alternatives?
   What would you do if the final set didn't contain any alternatives?
2. Does it make any difference in what order the constraints are applied to reach a determination of the final set? How about the effect of changing

the order of the constraints on the amount of time it takes to search the potential alternatives?

3. How could the multistage sampling apply to selecting a college? What would the grouping be? What major criteria would you apply to selecting a college?

4. Who might be influential in the selection of a college? Who really is the CDM in this choice? How does the existing system design come into play?

5. Contrast the approach used to develop this feasible set of alternatives, with the more typical approach of picking the same college where a person got the undergraduate degree.

Class Exercise 6-2

PERSONNEL JOB MATCHES

When one is about to graduate from school, one of the most frequent means of getting a job is through the classified ads of the newspaper.

A typical ad reads as follows for computer programmers.

Need: Programmers

Requires: (a) College degree in computer sciences
            (b) 2 years of COBOL programming on IBM 370
            (c) Good communication skills a must
            (d) Prefer a person who has project team ex-
                perience

1. In systems approach terms, what are these stated requirements (i.e., college degree in computer sciences)?

2. Why are these restrictions used? How solid are they? How definite are they really? Could a person with two years of college in the computer area actually do better on the job than a person who had a college degree?

3. What does good communication skills really mean? Could you state this requirement more specifically?

4. What is the difference in meaning between the language used in items (3) and (4)? That is, what do the words *must* and *prefer* imply?

5. Assume you were the personnel director for this company, and nobody applied for the job. What would you do? Assume 100 people sent in resumes and 70 of them meet the requirements. Now what would you do?

Class Exercise 6-3

WHERE TO HOLD A CONVENTION

Assume you were part of a selection committee whose task it was to select a city in which to hold the twenty-fifth annual convention of the XYZ Salesmen Association. The two-day conference is always held in the middle of January and has been attended by around 7000 people the past two years. The conference has always been held in the United States.

1. Why do people go to conferences? What considerations besides the agenda of the conference itself, are important?

2. What are the major components or variables in this systems design? That is, give a Chinese menu description of the various alternatives.

3. How many potential convention sites are there in the United States? How could you get the number down to a workable set?

4. Formulate five constraints that you would use. Name five major criteria that should be established.

5. Why is it that Las Vegas is more popular as a convention site than Cleveland? If a good convention site is found, why doesn't the association hold its conventions there every year?

Chapter Seven

How to Evaluate Alternatives

In chapters 5 and 6 we developed ways of designing system solutions and determined how to tell which solutions were in fact feasible. We also established how many feasible alternatives we would need to evaluate in depth and when necessary, how to reduce the alternatives to a limited number.

The basic question that will be addressed in this chapter is how to go about evaluating in depth the various feasible alternatives that have been developed and screened. To answer this question we need to first understand the basic structure of evaluating alternatives.

STRUCTURE OF ALTERNATIVE EVALUATION

We will build up the overall structure for evaluating any set of alternatives through a step-by-step process. To start off with, let's recall one of the basic systems diagrams.

In Fig. 7-1, some inputs are put into a black box. Within the box, some type of process takes place. The result of processing the inputs is a set of outputs.

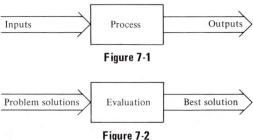

Figure 7-1

Figure 7-2

To this very basic, but general systems flow concept, we want to address our particular problem. That is, when we have more than one problem solution, how do we determine which is best?

As shown in Fig. 7-2, our various problem solutions are inputs to an *evaluation process.* The result or output of this process should be the best of the problem solutions that were inputted.

Let's get more specific about the evaluation process, which is diagrammed in Fig. 7-3. The problem solutions we are going to input to the evaluation process comprise a set of *workable solutions* that we have developed. For these various solutions, we would like to know their *relative worth.* That is, since all the solutions will solve the systems problem, we need to know which alternative does it best overall.

To make this type of determination, we need to look in more depth at the evaluation process. As shown in Fig. 7-3, the evaluation process is basically three steps or mini-processes. One process is to determine what the outcome or results will be if a particular alternative were adopted. Another process is to establish what does the underlying system actually need. The results of these two mini-processes are combined in a third process which attempts to determine the overall value of each workable solution. The resulting value of each alternative is then compared and the overall output of the evaluation process is the relative worth of the solutions.

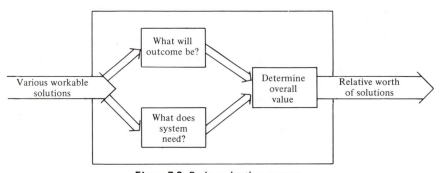

Figure 7-3 Basic evaluation process.

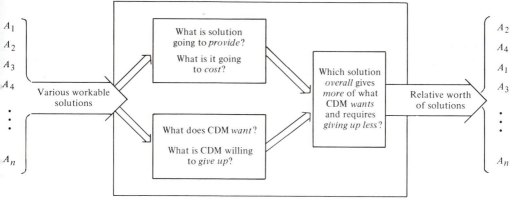

Figure 7-4 Evaluation process.

Let's delve even more deeply into the evaluation process as shown in Fig. 7-4.

Here the various workable solutions are now designated as alternative A_1, alternative A_2, through some last alternative A_N. These alternatives are inputted to the evaluation process. For each of the alternatives, we will need to determine what will the implementing of that solution provide, and further, what costs will be involved. Concurrently, we need to look at the systems context and determine what does the CDM want the system to do, and what is the CDM willing to give up to get certain results.

In Fig. 7-4, the outputs of these two mini-processes are combined to determine, of all the solutions, which solution overall gives more of what the CDM wants and requires giving up less. This evaluation results in a ranking of the relative worth of the solutions. As shown in the figure, alternative A_2 has the greatest relative worth, then alternative A_4, etc.

These concepts are related in Fig. 7-5 which gives an overview of an evaluation methodology. A much more detailed explanation will be given in chapter 12. In the overall structure of evaluating alternatives, we want to use as our inputs a set of feasible alternatives. For each of these alternatives, we will have to go through a *systems simulation* process to form a conjecture as to what the expected performance would be if each of these particular solutions were actually implemented in the system.

Further, from a study of the systems context, we will need to determine a set of criteria which we will use in developing the *systems utility function.* This function establishes the desired performance for this system over the planning horizon. The expected performance for each alternative is compared to the desirability of the various levels of performance within the *evaluation matrix.* The outcome of this matrix is an overall ranking on the relative desirability of each solution. This is shown in Fig. 7-5 as alternative A_2 most desirable and then alternative A_4, etc.

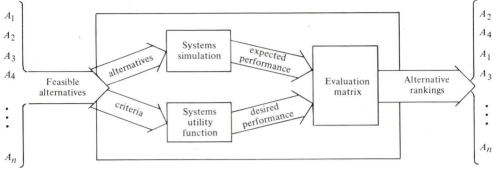

Figure 7-5 An evaluation methodology.

STRUCTURE OF EVALUATION EXAMPLES

Let's work through some very simplified examples in order to see how these concepts are all related. We will first cover the buying of a swimming pool using the least cost approach to evaluation. Moving up in complexity, we will then cover the purchase of a small business using the maximum profit approach. Lastly, the most complex evaluation using a multidimensional approach will be discussed in the example of purchasing a car.

Least Cost Evaluation

The first example concerns the selection of bids for building a specific swimming pool design. By keeping the pool design constant, we will measure the worth of each bid in terms of dollars. The alternatives are company A_1 which bid to build our specific pool design for $7800. Company A_2 bid $8200, etc. As shown in Fig. 7-6, there are four feasible bids. With each bid, we are expected to get a pool built to our specifications in terms of pool size, materials used, etc. The projected cost will vary with the bids. If we go with bid A_1, we expect to pay $7800, whereas bid A_4 will cost $9,900.

The desired performance is to get our pool for the least cost. According to this rule, the alternative bids are ranked in terms of desirability. Bid A_3 is most attractive, whereas bid A_4 is least. Note that the evaluation is made on the single dimension of dollar cost. Per se, no real tradeoffs are required if it is reasonable to assume all companies would build, in effect, the same pool.

Maximum Profit Evaluation

In the second example, the problem is to determine which is the best small business to buy. The feasible alternative businesses are called business A_1, business A_2, and business A_3. Profit is defined to be rev-

Problem: Determine which bid to accept for building a specific swimming pool design for your home.

Assume: Benefits, while in multiple dimensions, are held constant by specifying a particular pool design. All costs are in dollars.

Alternative solutions: Company A_1 = specific pool design for $7,800

Company A_2 = specific pool design for $8,200

Company A_3 = specific pool design for $6,300

Company A_4 = specific pool design for $9,900

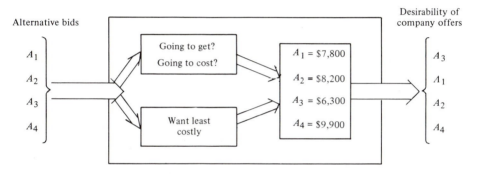

Figure 7-6 Swimming pool example. [*As evaluated:* Single dimension — least dollar cost; no real tradeoffs required.]

enues minus costs for the current year. Two further assumptions are necessary: (1) that we will use a single dimensional value system in which everything of importance can be measured in dollars; (2) that each business costs the same amount of money to purchase.

These points are delineated in Fig. 7-7. If we invest in business A_1, we expect to get revenues for the next year of $5000 and the costs will be $4000. So, for business A_1 the profit is expected to be $1000. Similar analysis for A_2 and A_3 yields profits of $3000 and $2300 respectively. These final overall results for each business alternative are compared and the alternatives ranked in terms of desirability. Thus, business A_2 with the largest expected profit of $3000 is most desirable and business A_1 with a profit of $1000 is least desired.

This example, while very much simplified, does give the basic concepts of the evaluation process. Note that while business A_1 is the least desired, it still is considered a workable solution. Further, all aspects of evaluating the worth of a small business other than revenues and costs as measured in dollars are considered unimportant.

Problem: Determine which small business to buy

Assume: Single dimension system where everything can be measured in dollars. All businesses cost the same to purchase.

Alternative solutions: Business A_1 Revenues will be $5,000
Costs will be $4,000

Business A_2 Revenues will be $6,000
Costs will be $3,000

Business A_3 Revenues will be $4,000
Costs will be $1,700

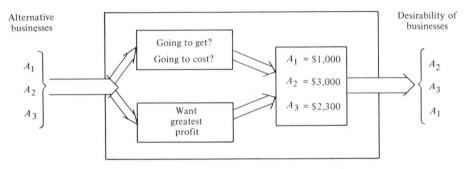

Figure 7-7 Business example. [*As evaluated:* All dimensions of evaluating a business other than dollars are considered unimportant.]

Multidimensional Evaluation

In our third example, we will analyze the problem in terms leading to a full discussion of the evaluation methodology in chapter 12.

The problem as shown in Fig. 7-8 is to determine what is the best car for the Jones family. Here the benefits and costs of the various alternative solutions will be measured in multiple dimensions (i.e., not just dollars). The alternative solutions are an overhaul of the existing system design which is a VW bug. A second alternative is a new '82 Reliant, etc. Each of the alternatives are feasible. They have met the constraints of less than $15,000, must get at least 15 mpg, etc.

Benefits for each car involve the good or desirable things it will provide. In this case, it would involve measurement on dimensions like status, looks, and dependability. Costs, on the other hand, are negative aspects implied with each alternative. These could be measured on dimensions such as initial cost, gas mileage, reliability, etc. In general, with aspects of the alternative which yield benefits, the more we can get, the better we like it. With those aspects which are costs, the less we have to face, the better we like it.

While it is not specified in Fig. 7-8, we would, for each alternative, determine what the status is, how attractive the car is, how dependable we feel it would be, etc. Additionally, we would check how much the initial purchase price is, what kind of gas mileage each car can be expected to get over the planning horizon, etc.

After all this information is gathered, the alternatives would be rated according to their performance with respect to the specified

Problem: Determine the best car for the Jones family.

Assume: Benefits and costs are multidimensional aspects.

Alternative solutions: Car A_1 = '73 VW (existing system – overhauled)

Car A_2 = '82 Reliant

Car A_3 = '82 Cutlass

Car A_4 = '82 Seville

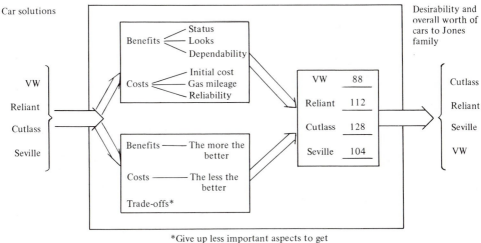

*Give up less important aspects to get more important considerations

Figure 7-8 Car example. [*As evaluated:* All relevant dimensions are considered.]

criteria. We will see in chapter 12 how to do this. But assume it has been done and the overall point results are shown in Fig. 7-8 as VW, 88 points; Reliant, 112 points; Cutlass, 138 points; and Seville, 104 points. The overall desirability of worth of the various cars for the Jones family would be Cutlass first, Reliant second, etc.

While it is clear that each of these three examples have been greatly simplified, they nevertheless illustrate the differences between the most used evaluation methods. Think through the methods that you have been taught in classes like accounting, finance, quantitative methods, management, psychology, etc., and see how they relate to the various methods above. In this text, we will be making a case for the multidimensional approach as the best overall.

SYSTEM OBJECTIVES

As was stated in chapter 4, problems are basically deviations between where a system is and where it is expected to be. To determine where a system should be, one needs to study the systems context.

From this study, one needs to establish what the key variables are and what their present state is. Additionally, what are the purposes of this system as determined by its relationship to the environment and what is the range of adaptability of the system? Derived from these thoughts and the CDM's perceptions of what is important, comes a set of *system objectives.*

> **System objectives** are desired standards of performance for the overall system.

System objectives are concerned with the overall aspects of a system (i.e., wholeness) rather than the specific subsystem. For example, in the case of a family, the question is not what is the best job for the husband, but rather what job should the husband take which is best for the family and considers his needs and wants. In most cases, the resulting job would not be the same considering both criteria.

There are generally multiple system objectives which are multidimensional and can't all be reduced to dollar cost. For example, General Electric has stated that it judges organizational performance not on the single goal of maximizing profits but by considering (1) profitability, (2) market position, (3) productivity, (4) product leadership, (5) personnel development, (6) employee attitudes, (7) public responsibility, and (8) balance between short-range and long-range goals (Horngren).

Note that these goals are multiple and potentially conflicting and cannot be explained away by saying they are all just ways of showing how to maximize profit in the long run anyway. Often, market share can only be enhanced by conceding lower profits. Responsibility to society and the company employees many times run counter to profit objectives. The drive to be No. 1 in the industry or to have product or service excellence can be self-defeating in a pure profit sense.

These same concepts apply in the selection of a place to live. This decision should involve the consideration of neighborhoods, quality of education, purchase price, aesthetics, size and layout of the home, etc. These aspects are multidimensional and can't nor should they all be resolved to dollar cost. This example, using the complete systematic systems approach is discussed in Chapter 17.

Objectives are usually conflicting. In the selection of an investment portfolio, those stocks which have the highest return also have the greatest risk. In a horse race, the horse with the least risk having odds like 2 to 1 pays much less to win than a high risk long-shot horse at 100 to 1. Thus, return and risk are conflicting objectives. One can't, in general, have both a high return and a low risk.

Steel corporations like Kaiser have conflicting objectives concerning profit and social responsibility. To be socially responsible by reducing pollution from the steel mills requires a large outlay of capital. Money spent for this reduces the profit Kaiser could make.

New cars delivered to California must have various anti-pollution devices, to help reduce the very serious smog problem in California. However, these devices lower gas mileage, add additional cost to the purchase price of the car, and increase the cars' maintenance cost due to valves burning out sooner, replacing anti-pollution devices, etc. Thus, while the reduction of smog is a very worthwhile objective, it conflicts with other objectives such as reducing the consumption of energy (i.e., lowering gas mileage increases the use of energy).

System objectives are usually stated very generally and are broad in scope. For example, a small suburban newspaper publishing company (i.e., the system) has a problem. What is the best typesetting process for the XYZ Newspaper Company? The objectives of the overall XYZ newspaper system could be as shown in Table 7-1.

The point of the examples is to emphasize that in the study of complex real-life systems such as businesses, governmental agencies, hospitals, educational institutions, churches, families, individuals, etc., that these cybernetic systems have multidimensional purposes which are both qualitative and quantitative in nature and much richer in scope than the usual one-dimensional objectives of maximize profit, minimize cost, or determine the best cost-effectiveness ratio.

Table 7-1 System objectives.

To increase the *desirability of produce* (newspaper) to *users.*

To decrease the *life cycle cost* of product (newspaper).

To increase the *reliability* of the *process.*

To enhance the growth *capability* of the firm.

To raise *employee morale* to the highest levels.

In summary, complex systems have multiple objectives which are conflicting and require system tradeoffs. Further, these objectives are multidimensional and shouldn't be evaluated strictly in dollar terms.

CRITERIA

The overall system objectives are broad in scope, very generally stated, and related to the overall system. What is needed is a refinement of the objectives which is tailored to the problem at hand. This needs to take place in the context of a system or a subsystem problem to enable the analyst to make a sound choice among alternative solutions. This is the function of criteria.

Criteria are more specific aspects of overall system objectives tailored to the system design choice.

To illustrate the relationship between system objectives and criteria, let's continue with the suburban newspaper example.

Thus, in trying to decide which of a variety of possible typesetting processes is best for this newspaper company, we would use criteria like

Table 7-2 Relationship objectives and criteria.

System Objectives	Criteria
Desirability of product	Versatility to customer Product quality
Life-cycle cost	Capital investment Page cost
Reliability of process	Personnel competence Maintenance of system
Growth capability	Time element Input volume Personnel competence
Employee morale	Status of personnel

versatility to customer and product quality as specific aspects of increasing the desirability of the newspaper to users. Capital investment and page cost would form the dimensions measured for life cycle cost.

This approach thus ensures that the criteria used in selecting a subsystem design will help to further the objectives of the overall system. However, these criteria must be defined. For example, the criterion of capital investment means many things to different people. Thus, we have to *define what we mean* when we say capital investment.

For this example,

Capital investment is defined as the initial purchase price of the typesetting equipment *or*

Capital investment is defined as the initial purchase price of the typesetting equipment plus five year annual operating expense minus ending salvage value *or*

Capital investment is defined as the initial purchase price of the typesetting equipment, additional building space required, and cost of training personnel

The point here is that no one definition will suffice in all situations for terms like profit, product quality, neighborhood, quality of education, salary, happiness, etc. We as analysts need to decide for each problem situation what are the best criteria to use and explicitly define what we mean by these terms.

As further background on these ideas, consider the criteria used by Talbott to select the best parachute for sky diving acrobatics (see Table 7-3). While the parachute example is relatively straightforward and lends itself to mainly quantitatively oriented criteria, it differs substantially from an example for selecting a life style shown in Table 7-4. This is a subset of the criteria developed by Sides. These criteria are much more qualitatively oriented and very complex. This is necessary when dealing with a subject such as life style which impacts on essentially all components of a family system.

HOW TO DETERMINE CRITERIA

In previous sections, we have discussed what systems objectives and criteria were and their relationship. We also gave specific examples to illustrate this very important aspect of the problem-solving process. But we need to go further by discussing the general categories of relevant criteria for a particular system.

Criteria Categories

We have stated in the section on the structure of alternative evaluation that cybernetic systems needed to be measured along multiple dimensions, — that all important aspects to evaluation should not be collapsed or forced into the single dimension of dollars. What other dimensions should be considered? To answer this question, we will define major criteria categories and give examples of the many dimensions that can be covered. What dimensions should be included in a particular systems problem and how to perform a multidimensional evaluation will be discussed in chapter 12.

We will define four major criteria categories, designated as technical, economic, psychological, and political.

Table 7-3 Acrobatic parachute criteria. [from Talbot]

Fit. The comfort which the harness and system design provides when adjusted to my body.

Cost. The total dollars which will be spent to purchase main and reserve canopies, containers, harnesses constituting a ready to jump system, excluding any freight or taxes.

Forward Speed. The maximum speed in miles per hour that the canopy will move relative to the ground in zero wind conditions in a "no brakes" configuration.

Rate of Descent on Landing. The minimum vertical velocity in feet per second which may be achieved at the point of touchdown.

Choice of Canopies. The number of differently designed main canopies which can be used interchangeably with no modification to any other components of the system.

Use for Relative Work. The degree of freedom of movement and safety which a particular system provides when on a relative workload.

Use for Jumpmastering. The degree of freedom of movement and safety the system would provide when handling students and static lines.

Packing Ease. The people and equipment which must be available in order to repack the main parachute per manufacturer instructions.

Forgivingness of Main Canopy. The ability of the jumper to make a radical turn or toggle movement without endangering his safety.

Safety of Reserve Procedure. The degree of safety inherent in the reserve procedure when the procedure must be used in a lifethreatening situation.

Number of Jumps Main Canopy Will Last. The number of jumps that will be put on a canopy before it is worn to the point of being unsafe.

The type of criteria in the technical category are defined as follows:

Technical criteria measure a systems outcome with respect to functional or operational characteristics usually expressed in quantitative terms.

The technical characteristics of a system are directly related to its primary purpose in being. Commonly considered to be objective in nature, technical criteria can be quantified, which refers to the ability to "put a number on it." Examples of this category would be miles-per-gallon, cubic feet of cargo space, debt/equity ratio, mean time between failure, size of house in square feet, etc. Also in this category are resources like time, space, manpower, etc.

After it has been determined that the alternative can perform the function desired, the next consideration is often cost. This criteria class will be called economic.

Economic criteria measure a systems outcome with respect to costs which are usually expressed in monetary terms.

This criterion is used in almost all evaluations in some form or other. It is primarily concerned with the dimension of dollars.

Table 7-4 Life-style criteria [adapted from Sides].

Freedom. *The degree of control over a member's actions and personal goals.* A member's ability to define what an objective is, when an objective is to be accomplished, and the method for accomplishment determines the level of life-style freedom. My ability to recognize the external influences in my life that alter the way my physiological, intellectual, and emotional structures deal with my ambitions and interests is an essential element of life-style freedom.

Standard of living. *The amount, quality, type, and method of accumulation of material objects.* Whether I am able to afford only the essentials necessary to maintaining a life style or am able to afford luxuries unrelated to its maintenance is a measure of life-style standard of living.

Challenge. *The degree to which there is a stimulating, exciting, motivating existence that encourages a member's growth and development.* Not a "one on one" confrontation, a life style offering challenge is a "one on himself" confrontation encouraging individual development. Identifying and solving changing problems, reaching into my mind and grasping new ideas and concepts, and feeling myself move with my environment are attributes of a challenging life style.

Location. *The extent of people, traffic, industry, congestion, and clean air affect the satisfaction derived from a life style's location.* The physical environment a member lives in must be conducive to his development.

Physical and mental health. *The amount of time that I have to read, to exercise and to think about myself.* A life style's ability to encourage a member to make positive corrective changes rather than submit to negative internal or external forces is a measure of a mentally healthy life style.

Social acceptability. *The degree to which society determines that a life style is detrimental, helpful, or has no effect on the maintenance of societal values.* While a member is evaluating his own life style, society will be judging it against traditionally accepted standards and mores.

Examples of criteria in this category would be price, life-cycle cost, down payment, capital investment, return on investment, etc.

The third criteria category is concerned with the psychological aspects of systems.

> **Psychological criteria** measure a systems outcome with respect to human characteristics usually expressed in qualitative terms.

The term qualitative is used in the usual sense of measuring non-quantitative things, that is, those aspects of a solution which can't be readily or meaningfully measured by "putting a number on it." Examples of this category would be beauty, happiness, safety, hassle, prestige, satisfaction, quality, independence, etc. These are commonly considered to be subjective in nature.

The last criteria category we will call political.

> **Political criteria** measure a systems outcome with respect to

power characteristics which are usually expressed in qualitative terms.

The word power is used in the sense of the factors necessary for insuring that a system solution will be endorsed initially and then later actually implemented in the system.

Examples would be risk, amount of change required, pressure, innovativeness of acceptance group, and environmental considerations like government, chain of command, etc. From these examples, it can be seen that the term political is used to measure aspects of the impact on various people-groups, i.e., users, decision-makers, unions, workers, etc., as to their acceptance of the problem-solution. This category may seem a little nebulous at this point, but a detailed discussion of its relevance and meaning will be given in chapter 13, "The Decision Process" and chapter 14, "Implementing the System Solution."

The last two criteria categories (i.e., psychological and political) aren't often used explicitly because they are generally considered to be irrational. Further complicating their use is the fact that they are subjective and don't readily lend themselves to measurement in precise quantitative terms.

Notwithstanding these comments, in the study of cybernetic systems these nonquantitative factors are generally as important, if not more so, than the easier-to-determine technical and economic criterion. Happiness is often more important than money, and it is as "rational" a consideration of systems performance as the functional criteria of work output. Making the necessary compromise in technical specifications to insure that the favored alternative will be implemented, is a "rational" consideration.

In this text, we are developing a way to handle system problems which must be viewed in complex, multidimensional schemes. This will be detailed and summarized in chapter 12.

Approaches

Given the criteria categories of *technical, economic, psychological,* and *political,* we have some idea of the range of things we should be looking for. But we haven't been shown how to determine which criteria are relevant for a particular systems problem. To do this, several approaches designated *explicit, implicit, derived,* and *comparison* will be covered.

Explicit: The most direct approach to determining what the systems objectives and criteria should be is to explicitly ask the CDM. This is generally what is suggested in textbooks and is very effective when it works. Unfortunately, it doesn't work very often for a variety of reasons. First, the CDM's are very busy people who generally don't have the time or the desire to work out in detail what they want done. Most decisionmakers just want the problem solved (i.e., the completed staff-work concept of chapter 10). Sometimes the CDM may know exactly what his objectives are in regard to a particular system, but he doesn't want to tell you. This could be because his goals aren't the same as the goals of the organization, or he may not feel you could understand the subtleties of what he is trying to accomplish, or he may not feel you will keep certain information in confidence, etc. Lastly, the CDM may not, in fact, know what he wants or desires. He may not know enough about the situation and its interrelationships to put forth his thoughts confidently. Additionally, it is sometimes not easy to analyze one's own wishes and specify exactly what one wants to accomplish.

Implicit: As we pointed out, explicitly asking the CDM doesn't always result in true nor meaningful criteria. A second approach is to study the existing system in some depth to determine its strengths and weaknesses. People are very familiar with the system they are a part of and almost always have very strong opinions on its desirable and un-desirable features.

By asking people with highly different perspectives about the ex-isting system, you should be able to implicitly determine the under-lying values of the organization or group. For this to be accurate, you would need, in the example of a business problem, to speak with people from different functional areas of the company (i.e., accounting, marketing, etc.) and on different levels (i.e., vice-presidents, workers, etc.) to insure you get different perspectives. Very few people see the whole picture.

One other method should be used, and that is to look at the ex-isting system where the system criteria can be verified. For a business example, that would be the budget. How is money actually allocated? How much money is really given to improve employee morale? What kind of manpower is given to quality control? This is a good check on what people say is important and what they actually do in practice. If the CDM and decision-makers say product quality is extremely im-portant, and yet allocate little resources for insuring that quality, then their talk is mostly rhetoric.

Derived: A third approach is to derive criteria from showing the CDM alternative designs. Because of all the difficulties with explicitly determining criteria and since people become locked into thinking in terms of the existing system capabilities, this method works backwards toward establishing objectives. The CDM and other decision-makers are shown new alternatives and asked whether they like the designs or not.

To be successful with this approach, the proposed system designs should be highly different — the theory being that one can't say why he or she likes or dislikes something without having some underlying values that she is comparing.

By comparing the characteristics of the alternatives, the ones the CDM liked with those he didn't, the analyst can derive the underlying values. If possible, the analyst should try to go even further by asking the CDM what he or she likes and dislikes about each alternative design. The derivation of the underlying values should be both quicker and more accurate if the analyst has the CDM's conclusions as to the advantages and disadvantages of each potential solution.

Comparison: The last approach is to look at similar systems. One needs to compare and contrast systems designs which are considered successful with those which are deemed unsuccessful. The analyst needs to establish how the two classes of designs differ and what are the most crucial aspects. This in turn implies which variables are keys to success or failure and they become major criteria. For example, if the systems problem were to select the best mate, one should look at marriages which are very successful and those which haven't been. From an analysis of this difference, some key criteria to use in selecting a mate become evident. Ann Landers suggests in her studies of divorce, the major reasons are first, sex; second, money; and third, parents-in-law. Contrast this thinking with the factors most people use in selecting a mate.

To illustrate the importance of systems objectives and criteria, consider the following story of how Michelangelo designed a sculpture for the Catholic church in the 16th century. The story is related in the book *The Agony and the Ecstasy* by Irving Stone.*

Michelangelo as a Systems Analyst

The pope summoned Michelangelo and told him he was commissioned to create a fine work of sculpture for the church. While Michelangelo was very pleased by this request, he wanted further clarification as to what kind of sculpture the pope wanted. The pope responded by blessing Michelangelo and wishing him well.

Michelangelo went back to his studio and thought about what kind of sculpture design would be best for the church. He finally decided to design a very large ten-figure sculpture. The figures would include life-size horses, men and women. Michelangelo, as a widely respected sculptor, knew this design would be recognized as a masterpiece.

*Adapted from *The Agony and The Ecstasy* by Irving Stone. Copyright © 1961 by Doubleday & Company, Inc. Reprinted by permission of Doubleday & Company, Inc.

So he proceeded to acquire the marble and began chiseling the stone. It took Michelangelo three years of constant labor to bring this to reality. It was indeed a masterpiece. He was now ready for the pope to see his work of art. A formal unveiling cermony was set, both for the pope to accept the sculpture on behalf of the Catholic church and also to approve payment to Michelangelo.

The pope came and looked at the design and told Michelangelo he didn't like it. Further, it wasn't what was wanted and there would be no payment until he created a design worthy of the church and the pope.

Michelangelo was crushed — three years of effort and the CDM (pope) didn't like the system solution (sculpture design). Michelangelo sulked in his room for months not knowing what to do. Finally he decided to ask for a meeting with the pope. Upon meeting with the church leader, he asked the pope what he wanted in the sculpture. The pope told him he didn't know, that Michelangelo was the designer and it was his job to determine a fine sculpturing piece.

Michelangelo was very frustrated with this nonanswer to his question. He didn't want to chance doing another sculpture and having the pope reject it. That approach not only hurt Michelangelo's ego, but it took three years to do a sculpture and additionally, he wasn't getting paid while doing his work. Further, this contract was Michelangelo's only real source of income.

So Michelangelo decided to use a different approach; he prepared four mini-model designs. These designs were complete, through just miniatures of what Michelangelo was proposing. Each design took only a few weeks to execute. He then showed the model designs to the pope and asked which ones he liked and why.

From this approach, Michelangelo derived what criteria the pope was implicitly using to determine a fine work of art for the church. Additionally, with a slight modification to one of the mini-designs, Michelangelo got the pope to agree that this was what he wanted.

Michelangelo then set out confidently to sculpture a full-size rendition of this particular model design. Thus, Michelangelo has not only saved himself much time and frustration, but he had gotten the CDM's (pope's) approval for the final piece of art.

Unfortunately for Michelangelo, during the three years it took to do the full sculpture, the pope died. A new pope (CDM) came and looked at the sculpture Michelangelo had almost completed and told him it was unacceptable. No money would be authorized until Michelangelo had produced a sculpture which reflected favorably on the church and the present pope.

Such is the agony of being a systems analyst!

RELATIONSHIP OF CRITERIA AND CONSTRAINTS

It is generally difficult for people, the first time they are exposed to the concepts of criteria and constraints, to explicitly understand the difference. So in this section, we will concentrate on the difference in purpose and in form.

In chapter 6, we established that constraints are restrictions or limitations which any system design must meet for it to be considered feasible. Using this idea, we also said constraints could be used as screening devices when a large number of potential solutions needed to be considered.

However, once a set of feasible alternatives has been generated, constraints are of no more value in determining which is the best solution. This is so, because by definition all feasible alternatives satisfy all constraints. Therefore, what is needed is a way to judge the potential performance of the feasible alternatives. This is the purpose of the criteria.

In addition to the difference in purpose of constraints and criteria, they differ in form. Constraints are dichotomous and criteria are continuous. That is, with constraints there is *no benefit gained by surpassing the conditions required.* This is a major difference between constraints and criteria. Constraints are either met or they are not. For example, a requirement of graduation from college is a 2.00 grade point average. It makes no difference as far as graduation eligibility whether one has a 2.01, 2.75, or 4.00 grade point average. The question asked with a constraint is, was it met or not? Is GPA greater than 2.00? If yes, then graduate. If no, graduation is impossible.

Further, say a person is trying to determine the best car to buy for herself. She has specified she will pick only from those cars which have air conditioning, have trunk space to hold at least four suitcases, and have smog devices installed which meet the 1982 standards. These are all constraints which limit the feasibility of various cars. Cars will either meet these constraints or not. Does the car have air conditioning? Yes or no. The fact that the trunk can hold six suitcases or that the smog devices are within the 1985 standards *is not important or worth considering further* in terms of the decision of what is the best car for that person. Suppose she has further decided that the cars she will pick from must get at least 20 miles per gallon and cost less than $7000. The various cars will either meet these conditions or not. Therefore, these conditions are constraints.

From the set of feasible cars, this person will decide the best car based on the overall results of considering total cost, appearance, roominess, safety, economy, etc. These are criteria. They are specific ways of determining the "best" car to buy, which is the overall purpose of the study.

Criteria differ from constraints in that constraints are dichotomous (do or don't meet) whereas criteria are continuous (the more the better if beneficial, the less the better if detrimental). Consider the criterion of total cost. All else being equal, the less the total dollar cost of the car the better. With the criterion of aesthetic appeal, the more beautiful the car looks to the CDM, the better she will like it.

The criterion of total cost could have an upper limit attached to it. The less the car costs, the better deal it is, all else held the same. But in no case will the CDM pay more than $7000 for a car. Therefore, even though it could be proven that a $34,000 Mercedes-Benz is the best car made, it is not an acceptable alternative to this CDM because the bank will not lend her more than $7000 for a car.

In this last example, total cost is used as both a criterion and a constraint. The restriction of $7000 is an external constraint which determines the feasible car alternatives. However, of all the cars which cost under $7000, the CDM prefers to pay the least (i.e., a $6000 car is preferred to a $6500 car and a $5800 car preferred to both). This preference is made with all other considerations (criteria/dimensions) held constant or assumed to be the same or equivalent. Thus, the concept of total cost is being used in different ways and for different purposes.

Class Exercise 7-1

DETERMINING YOUR VALUES

To get a feel for the concept of implicit criteria, do the following analysis of yourself as an individual system in determining your most important objectives or values.

1. State explicitly what you feel are the most important objectives in your life during the present time period.

2. Analyze how you have spent your money over the last three months. Consider your checking account statements, credit card purchases, cash outlays, etc. Determine the percentage of money spent in major categories like place to live, food, transportation, medical, entertainment, charity, education, etc.

3. Analyze how you spend your time. Over the last three months establish percentage of time spent on work, school, studying, traveling, watching TV, eating, exercise, sleeping, etc. Also consider these categories for fixed commitments and then look at what you do with your discretionary time.

4. From questions 2 and 3, generalize on what you implicitly are declaring as your most and least important objectives or values.

5. How do the explicit answers to objectives in question 1 differ from the implicit answers in question 4. Why do they differ? Which is more realistic?

6. How do these values relate to Maslow's hierarchy (1954)? If you find that you spend one-third of your life sleeping, does that imply you should spend one-third of your money on insuring you get a good night's sleep?

Class Exercise 7-2

WHAT IS THE BEST INVESTMENT?

Assume you are the controller for a medium-sized corporation, and you temporarily have excess cash on hand. Since a major objective of this firm is to maximize profit, you are considering the desirability of various ways to invest this money.

1. The proposed investment alternatives are:
   (a) keep money-on-hand
   (b) invest in short-term Treasury bills
   (c) invest in 6-month certificates of deposit at the local savings and loan.
       Suggest several other potential alternatives for using this money.
2. What are the advantages and disadvantages of each investment approach stated in question 1?
3. From the answers to question 2, derive the most relevant criteria that should be used in this company investment decision. How do these criteria differ from using return on investment (ROI) as the exclusive criterion?

Class Exercise 7-3

AIRLINE SELECTION CONSIDERATIONS

You are working as a travel agent. The secretary for the president of a major corporation has called you, and asked that a flight be booked from Los Angeles to Philadelphia on Wednesday, January 21st.

1. What major questions would you need to ask before you could determine the best airline flight? That is, establish the variables that airlines use to differentiate their product and make it more attractive to certain classes of passengers.
2. In your response to question 1, categorize these differences in terms of technical, economic, psychological, and political criteria.
3. Looking at the passenger situation from the airline's point of view, why do they offer such a wide range of discount plans with associated restrictions? What differences does it make to the airline that it costs about the same to fly the airplane whether there is one passenger aboard or two hundred and fifty?
4. After the DC-10 crash in Chicago in which several hundred passengers died, McDonnell-Douglas Aircraft Company spent millions of dollars in an advertising campaign to convince the public that the plane was a safe aircraft. Was this necessary to do? Why?

Chapter Eight

Judging Alternative Performance

In chapter 7, on the structure of alternative evaluation, we saw in Fig. 7-4 the need for learning what the CDM wants and what the CDM is willing to give up. To explain those points in more detail we covered the concepts of system objectives and criteria. In this chapter we need to address the questions What are the solutions going to provide? and What is it going to cost? These considerations will be concerned with the performance of the alternative solutions. Further, there is the need to establish the relative worth of performance as opposed to looking for absolutes. In this chapter, we will see how to establish a comparison base and establish a scale for performance.

PERFORMANCE

Basically, we need to measure the value of the resulting outcomes if a particular alternative solution were to be implemented. The general outcomes will be measured in terms of performance.

Performance is the expected outcome of a particular alternative if it were implemented.

Performance is generally considered in terms of *benefits* and *costs.*

Benefits are those positive aspects of an alternative solution. They are the advantages, the strong points, and the good elements in an alternative. They should definitely not be limited to those points which can be quantified, but should also include qualitative aspects.

Costs are those negative aspects of an alternative solution. They are the disadvantages, the weak points, or the bad elements in an alternative. They should definitely not be limited to dollar costs but should include qualitative aspects as well as other quantitative points.

Using the definitions of performance, benefits, and costs, we can see that if the expected outcome of a particular solution yields "good" things, this is considered beneficial performance. Conversely, if the expected outcome requires "bad" things, this is considered negative performance, Here "things" are the aspects of the solution which relate to the overall system objectives as measured by the more specific criteria.

The general relationship between performance and criteria can take basically one of two general shapes. If the criteria measure something which is desirable or beneficial for the system, then the CDM and others want (1) the more the better and (2) the most they can get. This relationship is depicted in Fig. 8-1.

The curve shows that the greater the performance (i.e., movement to the right), the greater the desirability (i.e., movement upward). Additionally, note that there is no maximum. Therefore the CDM will want as much as he can get.

If the criteria measure something which is not desirable or considered negative or costly to the system, then the CDM and others want (1) the less the better and (2) the least they have to take. Further, be-

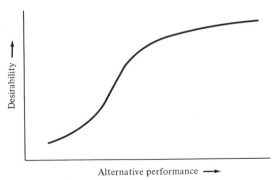

Figure 8-1   **Beneficial criteria.**

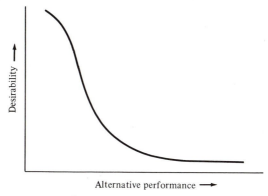

Figure 8-2 Costly criteria.

cause there is no minimum, the CDM will strive for as little as possible. This relationship is shown in Fig. 8-2.

Let's consider some examples of beneficial criteria and costly criteria. In the case of buying a car, a beneficial criterion would be gas mileage. From Fig. 8-3, we can see that the more miles-per-gallon a car gets, the more attractive it is to the CDM. While the exact shape of the curve would vary among CDM's and their particular system contexts, this general shape holds. Note that this curve is plotted while holding all other aspects of an alternative constant. You look at one dimension at a time and then at the end, put it all together. If at this point in the analysis the CDM concerned herself with the purchase price or interior size of the various cars which give 20 mpg or 40 mpg, the person confronts the evaluation too early in the problem-solving cycle and tends to confuse the issues.

As an example of costly criteria performance, consider the purchase price of the cars. With all other aspects of alternative performance

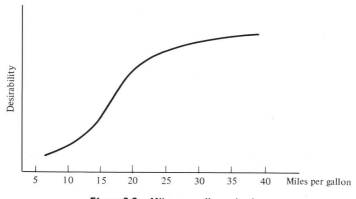

Figure 8-3 Miles-per-gallon criteria.

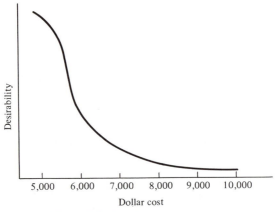

Figure 8-4 Purchase price criteria.

held constant, the general shape of the curve is shown in Fig. 8-4. According to the figure, the CDM would prefer to pay $8000 for a car as compared to $10,000 (everything else considered equal). Further, a purchase price of $5800 is more desirable than $8000. Thus, on a negative or costly criteria the less the better. There is no minimum in the sense the CDM would like to pay nothing. In fact, he would prefer to be paid for buying a car. Similarly, in the mpg example, the CDM would like to have a car which gets 100 mpg; in fact, the best would be one that runs on water, etc.

As examples of more qualitative criteria, consider happiness and hassle. Happiness is a personal system criterion in which the more the better is preferred (i.e., beneficial), whereas hassle is a costly criterion in which the least possible is desirable.

RELATIVE WORTH OF PERFORMANCE

We have just defined performance as it relates in a general way to beneficial and costly criteria. But we need to get much more specific so that we can evaluate the relative worth of performance of various alternatives. We have to be able to do this since we have seen that we will need to deal with multiple solutions to a systems problem.

Further, we have learned that there are no absolutes in evaluating performance. While "the more the better" is a general rule in evaluating performance of beneficial criteria, it doesn't tell us what is good performance nor what is bad performance. Although 150 mpg is desired and certainly preferable to 25 mpg, is it technically possible? Further, is it reasonable to expect such cars to be available during the planning horizon? In other words, to be useful the evaluation of performance needs to be grounded in the reality of the times.

To be sure we are facing reality, we should define good and bad performance in relationship to that performance which can usually be expected.

Good performance refers to those outcomes which give better than the normal or usual performance in a particular situation.

Bad performance refers to those outcomes which give less than the normal or usual performance in a particular situation.

In the case of the mpg criteria of the car example, assume cars usually get 20 mpg. Then a car which gets 30 mpg would be considered as giving good performance, whereas in the same setting, a car which gets 15 mpg would be giving bad performance. However, this same car which gets 15 mpg in a different setting where the norm was 10 mpg would be considered as giving good performance.

This example points up two very important aspects involved in evaluating the relative worth of alternatives. First, absolute performance isn't what is critical; rather it is relative performance. A car which gets 15 mpg cannot be classified as getting good or bad performance until there is a definition of what cars usually get in mpg. Secondly, there is both good and bad performance with respect to both beneficial and costly criteria. For example, with the beneficial criteria the more mpg the better. If the norm is 30 mpg than 40 mpg is good and 20 mpg is bad performance. Similarly, with the costly criterion of purchase price, if the norm is $8000 then a $7000 price is considered good, whereas, a $10,000 price is considered bad. See Figs. 8-5 and 8-6.

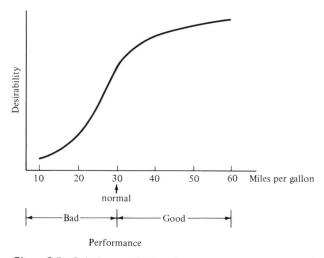

Figure 8-5 *Relative worth of performance:* beneficial criteria.

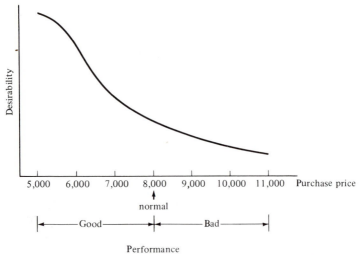

Figure 8-6 *Relative worth of performance:* costly criteria.

DEVELOPING A COMPARISON BASE

In the previous discussion on the relative worth of performance, it was stated that the determination of good or bad performance should be made in relationship to what is the usual or normal situation. But we didn't show how to determine the usual situation.

Set of Comparables

In real estate when one goes to buy a house or arrange for bank refinancing, there is a need to establish the value of the home in question. It is the job of the appraisers in many cases to answer this question for the bank. Appraisers use a variety of methods to come up with an answer, including the *accounting* approach, (i.e., cost of re-building the home at today's prices); *value-added*, (i.e., initial buying price plus the improvements you have made); and lastly, *market value.* The market value approach basically asks the question, What have similar houses been selling for most recently? To this the appraiser or real estate agent looks at sales of homes over the last three or four months in the same neighborhood of the house being appraised. Further, it is most desirable that the comparable homes be similar as to the number of bedrooms, square footage of house and lot, age of the home, view, pool, etc.,

The *set of comparables* approach basically determines the value of something by determining what value has been established for similar items. In this regard, it tends to be a much more useful approach than either the accounting approach or value-added. These other approaches tend to be used when the item being valued is very unique (i.e., no real set of comparables).

In any case, we will use as our main method for establishing the normal performance, the method of set of comparables. Let's see how this would work with our car example as it relates to mpg. Assume you are looking at a car which gets 25 mpg. Is this good or bad? Compared to what? Well, it should be compared to the mpg that all similar cars would get.

Assume you were looking only at new cars and you were going to use the Environmental Protection Agency's gas mileage estimates to develop your comparison base. After looking at say 100 different new car models, you arrived at the results for gas mileage as represented graphically in Fig. 8-7.

Most likely you would get a curve which is shaped like a bell. This is called a normal distribution curve. The greatest number of cars would have mileage around the peak of the curve which in this case would be 20 mpg. A small percentage of cars would get 10 mpg or less. Similarly, a small percentage of cars would get 35 mpg or greater.

Given this data on the performance of a set of comparable cars on mpg, we can state that 25 mpg would be good since it is greater than the usual 20 mpg. On the other hand, 15 mpg would be poor.

Of those 100 cars that we took as a set of comparables, we required only that they be new cars. So our set had Lincolns, Hondas, Ford vans, etc. If we were considering buying just subcompact cars, then it wouldn't make much sense to have Lincolns, Ford vans, etc., in the comparison base. We recall that to be useful, the comparison has to be of similar things. So in this case, it would make sense to limit the comparison cars to new subcompacts like Honda, Toyota, Pinto, etc. Assume we did so for 100 different new cars and got the distribution curve for mpg shown in Fig. 8-8.

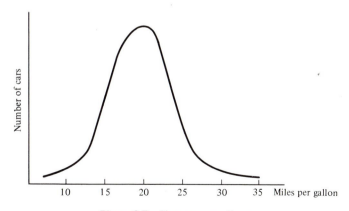

Figure 8-7   New car gas mileage.

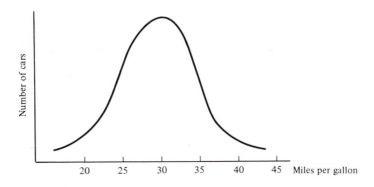

Figure 8-8 New subcompact cars.

Here we note that the average or normal is 30 mpg. We get 30 mpg by using the EPA sticker estimate for each of the 100 cars and then compute the average figure.

In discussing Fig. 8-7, we stated that a particular new subcompact car we were looking at got 25 mpg. This was considered good performance. But this same car using the comparison base of Fig. 8-8 would be considered having bad performance since 25 mpg is less than the average or normal 30 mpg that new subcompact cars get.

The crucial difference lies in the relevance of the set of comparables picked. In the first case (Fig. 8-7), the comparison base was all new cars to include vans, large cars, subcompacts, etc., so it was biased toward making the little car look good. Whereas in Fig. 8-8 the comparison base was limited to similar subcompacts.

The general rule is that the set of comparables should conform as closely as possible to the characteristics of the alternative being valued. When looking at subcompacts, the comparison base should be subcompacts. However, if one were looking at a Honda, and a Chevy van, and a VW bus, then the comparison base should have a similar mix of vehicles or be from the same "class" of vehicles.

Determining Appropriate Sets

To get a useful set of comparables depends on establishing the appropriate class of alternatives upon which to measure typical ranges of performance. A good way to determine the class of solutions is to remember that they need to be related to the systems context and the particular problem at hand.

We stated earlier that constraints are a very useful means of establishing the difference between potential and feasible alternatives. What we want to really know is what kind of performance can be expected from the total set of feasible alternatives. Thus, the set of comparables is the total set of feasible alternatives.

To see both the why and the how of determining the appropriate range of performance, consider the example of buying a house. Let's take the relatively straightforward criterion of size of house to look at in depth. Assume size of house means the living space contained within the house itself, excluding the garage, as measured in square feet. The general shape of the desirability performance curve is shown in Fig. 8-9.

As shown, the CDM prefers to have a 1600 square foot instead of a 1200 square foot home, everything else held equal. Further, she most desires a 2800 square foot home or even larger if she can get it.

To the curve of Fig. 8-9, let's now add the general selling price of homes for the various sizes. This curve shows that homes in the $50,000 price bracket generally have between 900 and 1300 square feet. If you want a home with over 3000 square feet, you will have to pay around $500,000.

The first diagram (Fig. 8-9) gave the CDM's general desires for various size homes. While this background knowledge is helpful, it doesn't really help determine which alternative to finally select. We have not distinguished dreams from reality. We need to ask what is the size range of houses that the CDM can realistically afford? That is, what are the financial constraints?

We see why these considerations are necessary in Fig. 8-10. If a person can only afford a $50,000 home, it is a waste of time to consider or look at $500,000 homes. Remember, one of the fundamental premises of the systems approach was to make the system under study more effective. To do this, one needs to work in reality, no matter how distasteful this might be in some situations. Dreaming is wonderful and inspiring but it is not problem-solving!

Returning to Fig. 8-10, assume that financial analysis of a person's income and present savings indicates she could afford up to a $70,000 home. Then, according to Fig. 8-10, the appropriate range for size of house would be 800-1600 square feet. The upper limit is 1600 square

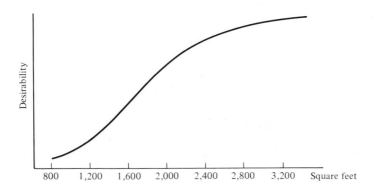

Figure 8-9   Size of house criteria.

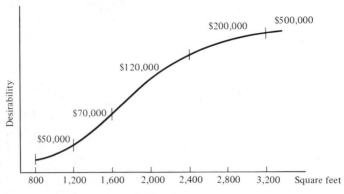

Figure 8-10 Size of house.

feet and the lower is 800 square feet. Presumably, no house could be found under 800 square feet.

For another couple, their income is such that they can finance a $500,000 home. The size range is then 800 to 3200 square feet. However, we have considered only one constraint, income. In actuality, we would need to consider other restrictions. This could include items like (1) house must be located in Beverly Hills; (2) house must have five bedrooms; (3) home must have a swimming pool, etc. Considering these constraints, the appropriate range for this couple could be 2800-3200 square feet.

As we can see from this example, what is the appropriate range for one situation can be highly different for another system. That is, while a 1600 square foot home is very large for a couple who can only afford up to $70,000 homes, it is so small to be unacceptable to a family who can afford $500,000 homes.

In summary, developing a comparison base involves determining a set of comparables. This set to be useful must match closely the characteristics of the alternatives to be evaluated. Constraints offer the best way of determining the appropriate class of alternatives, by removing nonfeasible solutions. The set of feasible alternatives is the comparison base.

SCALES OF PERFORMANCE

Up to this point, the general idea of performance has been discussed as it relates to criteria and system objectives. We have considered how to establish the appropriate range of performance, and developed a formal definition of the concept of good and bad performance. We now need to integrate all these concepts by giving a more

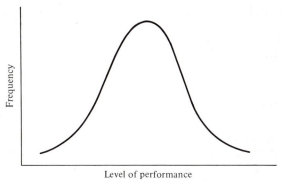

Level of performance

Figure 8-11 Level of performance curve.

formalized definition of level of performance and its relation to systems utility.

Classification of Performance Levels

Levels of performance may be conceptually defined in terms of a normal curve (see Fig. 8-11). This curve represents the most prevalent distribution of performance. It is symmetrically bell-shaped and has specific mathematical characteristics in terms of standard deviation, etc. We won't get this detailed; rather we just want to utilize the underlying concept.

Using this normal curve, we will define certain levels of performance as shown in Fig. 8-12.

Exceptional performance is the level of performance indicated by the top 10% or 90 percentile. This information is well beyond

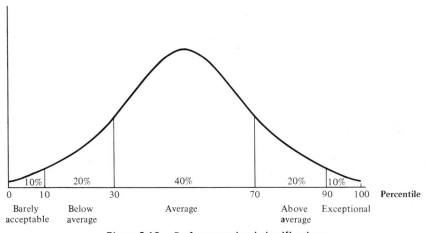

Figure 8-12 Performance level classifications.

what could be normally expected from systems designs in this category.

Above average performance is the level of performance indicated by the 70 to 90 percentile. This performance is beyond what could be normally expected from systems designs in this category.

Average performance is the level of performance indicated by the 30 to 70 percentile. This performance is what could be normally expected from systems designs in this category.

Below average performance is the level of performance indicated by the 10 to 30 percentile. This performance is below what could normally be expected.

Barely acceptable performance is the level of performance indicated by the lowest 10% or up to the 10 percentile. This performance is well below what could normally be expected.

Point Value of Performance

In a formal evaluation (see chapter 12), we will find it useful to express the classifications of exceptional, average, etc. as numeric values. To do this, we will use a 10-point rating scale. This is an arbitrary choice. The scale could be 20, 100, 1000 or any other number of points or divisions. In practice, the 10-point scale seems to work out well so we will use it consistently throughout this text.

The underlying performance curve is continuous, even though we have grouped the curve into five categories. The 10-point scale is continuous also and the points are assigned as shown. Five points are given for average performance, 9 points for exceptional and 1 point for barely acceptable performance. Thus, 9 points is given for the "average" of the exceptional range, 8 points for "high" above-average and low-exceptional, etc. Beyond getting the general flavor here, it isn't worth dissecting the categories too finely.

These concepts will be discussed in more detail in chapter 12, but to highlight these thoughts, let's relate Fig. 8-13 to the assignment of grades. All through your educational experience you have been rated on your performances. At the end of the quarter, semester, or school year, each teacher gave you an A, B, C, D, or F grade.

How was this grade determined and what impact did your fellow students have on the value of your performance? Teachers rarely have absolute standards which they apply irrespective of the background of the students in a particular classroom. More generally, teachers over the

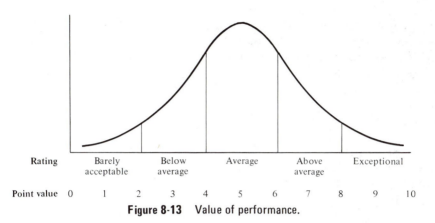

Rating	Barely acceptable	Below average	Average	Above average	Exceptional

| Point value | 0 | 1 | 2 | 3 | 4 | 5 | 6 | 7 | 8 | 9 | 10 |

Figure 8-13 Value of performance.

years develop a sense of what students can generally do in a particular class at a particular grade level, i.e., 9th grade algebra, college-seniors in a computer class, graduate MBA students in a management class, etc.

From this, they develop a scheme of performance ratings for a set of comparables (students). If you do work which is well beyond the norm and generally in the top 10%, you have shown exceptional performance and get an A. If you work very hard, but your resulting performance is less than what can normally be expected for students in a particular class, you receive a D.

Performance Measures

We have developed to this point criteria for evaluating feasible alternatives, and demonstrated a comparison base approach to establishing the relative value of performance. We then defined a classification scheme for various performance levels and converted its relationships to a 10-point scale.

What remains is the need for defining performance measures.

> **Performance measures** are yardsticks which explicitly show in quantitative and qualitative terms what is barely acceptable, below average, average, above average, and exceptional systems performance.

As an example to clarify what has been said about performance measures, consider one of the criteria that General Electric uses for judging organizational performance—profitability. Assume a corporation has set up a form of decentralized control which is very popular these days called profit centers. If one accepts the thinking that companies function best only when they maximize profit, then the success of various profit centers is very easy to evaluate. The greater

133

the profit, the better the performance, period! Even here, however, the operational definition of what profitability means, as well as specific guidelines for measuring barely acceptable through exceptional performance, will prove useful to all concerned.

How does one go about specifying criteria and performance measures? For the profit center example, one of the first things to do is to decide on a definition of profitability. In this case assume it is decided that the measure of return on investment is the most appropriate index of profitability. Note that the exact way ROI would be determined should also be spelled out in detail. It must be further determined what is barely acceptable profit performance and what is exceptional performance. This actual measure of performance is not a general thing but is tailored to the particular system one is looking at. The analyst may look at the historical performance of this division, company, or branch and see what levels of return have been evident in the last five to ten years. He could check how the industry does overall and also look at the performance of competitor's divisions. Suppose these checks show that the historical ROI for this division has been 22 percent and the industry average is 20 percent. After discussion with top management and profit center leaders, a yardstick of ROI could be developed as shown in Table 8-1.

Using this figure, the profit center manager now knows that if the center has a ROI profit of 23 percent, he is doing above average work. If the center ROI is 10 percent, he would best go look for another job.

In summary, what the analyst is eventually working toward is the evaluation of various feasible alternative solutions. How will the analyst determine which alternative is best overall for the system being studied? He will need to see how each alternative does for each criterion specified. Great, but in the case of General Electric, how does one decide what acceptable public responsibility is? Good public responsibility does not mean the same thing to everyone (i.e., to General Electric's president, Ralph Nader, a radical, or a stockholder). How are middle managers to know what the top management of the company really wants concerning public responsibility? Is it just lip service or is it worth trading off profits? How much profit must be returned before heads will roll?

In other words, to help clarify what is meant by good public re-

Table 8-1 Levels of performance.

Criteria	Barely Acceptable	Below Avg.	Average	Above Avg.	Exceptional
ROI	15%	17%	19%	22%	28%

sponsibility or any other criterion, the development of some kind of measures of performance is necessary. These measures can be implicit—which leads to a wide range of interpretations—or they can be made explicit—which leads to a more uniform understanding among a wider audience of what constitutes good and poor performance.

Theoretically, after the yardsticks have been defined and the measurements specified in operational terms, then anyone could measure how well a system would perform under various alternative designs and all would come to the same conclusion. Note, however this ideal is generally way too costly and limiting to reach in actual practice, although the general concept holds.

Class Exercise 8-1

RELATIVE PERFORMANCE

To see how the concept of a set of comparables and relative performance relates to everyday experience, answer the following questions.

1. What is a good price for a gallon of gasoline? Does it make any difference if the time frame is the 1900's, 1970's or 1980's? Why does it or doesn't it?

2. What height does it take to be considered a tall person? In professional basketball, they speak of a man who is 6'9" tall, who plays the center position, as very small. How can this be?

3. Around thirty years ago, a man first ran the mile in less than four minutes—an unbelievable accomplishment. Today, you would have to run under 3 minutes 50 seconds to even be considered good, let alone unbelievable. A four-minute mile is considered high school stuff. In sports, how are standards of performance set? Why is something considered unbelievable one year and ho-hum several years later?

4. How does the Internal Revenue Service decide which tax returns to audit? That is, they process 100 million returns, but only have the staff to thoroughly check 50,000 returns. Which returns should they review? How would they decide, a priori, what a reasonable or unreasonable amount is for deductions? Would this figure vary by income class? Could the IRS establish a set of comparables and determine a normal distribution of deductions? Or should they just audit those individuals whose returns were in the top 10% of all those who took deductions?

Class Exercise 8-2

COMPUTER TECHNOLOGY

1. In computer technology, the price of hardware is decreasing each year while the performance is improving. Desk-top computers today, which sell for several thousand dollars, have performance which exceeds that of com-

puters which cost around a million dollars twenty years ago. How does technology impact the definition of good and bad performance. What is the purpose of technological advancements?

2. How should one develop a set of comparables for classifying computer performance? Is it fair or useful to compare personal computers (i.e., Radio Shack), minicomputers (i.e., Digital Equipment) and mainframe (i.e., IBM) computers in terms of relative performance? How does a specific company's needs affect this choice?

3. In the study of some two hundred small business computers listed in *DATAPRO,* I found the following ranges of technical characteristics:

Characteristic	Bottom 10%	Average 40-60%	Top 10%
Word length	8-bit	16-bit	32-bit
Internal memory	16K	32K	64K
Cycle rate (kHz)	200	750	1500
Higher level languages	1	2	5
Firmware (%)	10	20	50

Given this relative performance on a set of comparables, what implication does this have in evaluating various small business computer alternatives? How does this set of comparables differ from say five actual computers that I would evaluate in the final set of choices? Could all five of these computer choices be in the top 10% on a particular characteristic and thus all rate as 9's or 10's?

4. What is the relationship between technical characteristics or performance and criteria; Why can't *DATAPRO* state which is the best computer to buy?

Class Exercise 8-3

HOW MUCH ARE YOU WORTH?

You are working toward getting a college degree. After you have graduated, you will face a major decision point in your life—choosing a job in your new career field.

1. One of the major criteria should be initial salary. How are you going to determine what you are worth in terms of salary to potential employers?

2. How useful would salary surveys be for your career field, as published annually by a nationwide employment service? How about a survey by your local college's placement service of the latest 100 students who accepted employment offers?

3. Is a good salary offer dependent on what your major is? That is, does it matter whether you are a chemical engineer, programmer/analyst, or history major? Why? How can this be, when it costs a person the same amount of time and tuition to get each of these degrees?

4. Does it make any difference where you went to college? How about your grade point average?

5. How can a $30,000 a year offer look so good to you, when Pete Rose of professional baseball is disgruntled with a $500,000 a year offer?

6. What is the relationship between your worth as an individual and the salary an employer is willing to pay?

Chapter Nine

Gathering Information

In the course of doing a formal system study, the analyst normally has to gather much information. Unfortunately, there is never enough time to gather all the information needed. Thus the analyst needs to know how to search for information and secondly, he needs to understand the basic methods of gathering information and their relative strengths and weaknesses.

HOW TO SEARCH FOR INFORMATION

There are many different strategies for gathering the information that will be needed in solving a systems problem. These strategies range from jumping in and gathering "all the facts" in the greatest detail possible, to being highly selective in gathering information but proceeding in a step-by-step fashion. The approach recommended in this text is called the *adaptive search procedure.*

Attuned to Study Resources

The adaptive search procedure is based on the consideration that the search for information should be tuned to the time, manpower, and dollars granted for the system study. As we will see in chapter 10, these study resources should be related to the problem importance and the *time to decision.*

The greater the importance of the problem, the more study resources should be applied. However, this is tempered by the deadline when a decision must be made. In all cases, the study results must be available to the CDM prior to that time, in order to have a bearing on the outcome.

A common fault among systems analysts is to get so immersed in the problem and trying to come up with the perfect (i.e., optimal) solution, that they forget the real purpose of doing a system study— that is, to provide the CDM a better solution or at least a more confident choice than he would have gotten had he not commissioned the study.

This can't be done if the analyst comes up with a recommended alternative after the CDM had to make a decision. So the search procedures must be flexible enough to show the analyst how to proceed to gather information whether the time available is 15 minutes, 1 day, 3 months, or 5 years.

With a little thought it is clear one can't gather the same amount of information if one has 15 minutes to do a study versus 1 year. So something has to give. One can't gather all facts, can't check out all alternatives, etc., and therefore we need a method which is selective and indicates to us where to look and to what depth to conduct research.

Selective

Initially when assigned a problem, the analyst needs to try to get an overall perspective of what the situation is. Therefore, information should be gathered about the various aspects of the study—quickly and not in much depth. The purpose is to establish a general framework or overview of how things fit together. At this point there is no effort to establish what the problem really is or what is causing it or what the solutions are—just generally how elements fit together in the problem area.

After this initial look at the problem, the analyst should have enough background on the overall situation so he can profitably go back to management (CDM) and discuss their initial guidance of what they felt the problem really was and also clarify their thinking in relationship to what he has learned so far.

Now one is ready to go look more penetratingly into what is the real problem and what are some of the basic causes (relationships) behind it. Because the analyst knows that he is not going to be able to gather all possible information, it is very important to determine which areas to look into and which to forget (at least initially). Investigate first those areas which look most promising. That is, those areas which look like they have a major impact on the problem and which there is also a good chance of affecting.

Since one can't or doesn't want to cover all aspects of a system in the fullest detail, one needs to distinguish the most relevant concerns from those which are unimportant. We have seen how the use of constraints is such a technique for screening potential alternatives. Thus to get to a set of alternatives which is worth our time to study and evaluate in more depth, we need to concern ourselves with aspects that the CDM has control over. Those aspects which he doesn't control should be taken as givens.

Highly affectable areas are those situations where in essence the CDM has control (i.e., where things can be changed if it appears necessary). It is better to concentrate initially on these areas and avoid those which are externally controlled and which will be much harder if not impossible to change. For example, assume you were asked by the newly elected President of the United States to formulate a plan that will "get this country moving" economically. You would soon find out that even though the Federal budget approved by the President is hundreds of billion dollars, the President in reality only has direct say over some 20 percent of it. The rest is written into law and would be very difficult to get changed (i.e., Social Security benefits). Therefore, the initial probes should cover those areas which the President (CDM) can influence directly.

The other major factor which would influence where to initially search would be those areas which are directly related to dominant criteria. This may be called "the squeaky wheel concept." If for example the clues provided by management or others concerning falling company sales is that product costs are too high, or product quality is too low, or that service is not reliable enough in relationship to our competitors, then the analyst can start his initial search in these areas to see if, in fact, this is true and why it is happening.

After identifying some initial search areas, it is important to determine how deeply these areas should be probed. We shouldn't get bogged down into too much detail at this point until we have established that in fact these areas are useful to the problem solution. We will change search areas as we get new information. Since it is most likely that we are working in uncertainty or partial ignorance, we cannot be sure we are searching in the best places. We can try some areas and if they don't work out or are not returning enough for our time, or we uncover new information which leads us elsewhere, we'll pick new areas. As the study goes along we should become able to make better decisions on areas to investigate because of what we have learned so far in our research.

Iterative

In doing a system study, there are two basic philosophical approaches to how to solve problems. These methods can be categorized as *linear* and *iterative*. The linear approach is a straightforward, step-by-step plan for attaining the problem solution. The iterative approach has the same overall purpose of the linear method, but covers the steps in a cyclical way, which unfolds toward problem solutions (see Fig. 9-1).

The numbers within each box are meant to specify the usually stated steps in problem solving. These are: (1) state the problem; (2) determine the objectives; (3) gather available facts; (4) determine the solution; and (5) implement the solution.

In the linear procedure, each step is taken in order. Moreover, one doesn't normally proceed to the next step until the previous step is completed. For example, one must know exactly what the problem is before he should attempt to determine objectives or come up with solutions. Finally if one does a thorough job in the earlier phases, he will be better able to determine *the* solution, i.e., the correct answer to the problem.

In the iterative approach, five stages have also been indicated to represent roughly the same five steps as in the linear approach. However, the steps aren't necessarily taken in order, or even completed before proceeding to a subsequent step. Furthermore, the steps are continually reexamined and the results changed, modified, or kept the

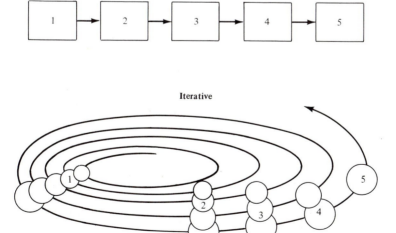

Figure 9-1 Philosophical approach to a system study.

same. It is not the intent of the iterative approach to determine *the* correct solution to a problem. Rather it is assumed that there are multiple solutions which could and should be determined and compared. Additionally, consideration of the later steps may help clarify what one is or should be doing in earlier steps. One last point: all relevant facts can't be determined and even if they could it most likely isn't worth the cost to do so.

In summary, the linear and iterative approaches differ significantly in their fundamental assumptions. The linear method is widely espoused and considered by many to be the best procedure to follow for problem-solving. For the purposes of systems analysis, however, we will find the iterative approach to be a much more effective procedure.

Systematic

The adaptive search procedure as we have seen, is selective in approach and attuned to the study resources available. Furthermore, it is iterative in nature. There is a continual recycling of the major aspects of a study, which becomes increasingly penetrating as it spirals toward a conclusion.

But to make this overall procedure most useful, there has to be a systematic way of implementing these concepts. The analyst will be guided in his *systems study* by completing a series of three progress reports—*preliminary, feasibility,* and *evaluation.* Each report documents a particular phase of the system study.

> **Preliminary report** is prepared to ascertain what the problem really is and to determine what steps should be taken to resolve it.

> **Feasibility report** is prepared to discover if there are any workable solutions to the stated problem.

> **Evaluation report** is prepared to assess the overall desirability of the alternative system designs that have been advanced.

Each report builds successively on the previous one. For example, the feasibility report includes all the elements that are in the preliminary report plus several major new items. The evaluation report includes all the elements in the feasibility report plus several major new items, etc. However, it is important to recognize that the initial items are restudied and formulated at a deeper level of understanding with each iteration.

The three reports lead to the completion of the final report whose overall purpose is:

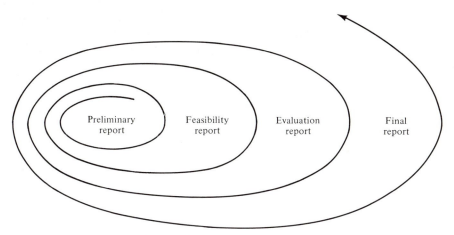

Figure 9-2 Systems study reports.

Final report is to determine the best way to handle the problem identified, and in so doing increase the overall effectiveness of the system.

This is conceptually shown in Fig. 9-2 and is explained in much more depth in chapter 10 (Fig. 10-4).

In summary of the methodology for gathering information, we have put forth the adaptive search procedure as the primary approach used in this text. Its major characteristics for guiding the analyst to the determination of a good solution to the systems problem are listed in Table 9-1.

Table 9-1 Adaptive search procedure.

Attuned to study resources

Selective information gathering

Iterative and increasingly more penetrating

Systematic

BASIC METHODS OF GATHERING INFORMATION

In the course of solving complex systems problems, the analyst needs to gather much information regarding the system and its context, establishing what the problem is, developing objectives, designing and evaluating alternatives, communicating the study results, implementing the preferred solution, and performing a systems audit. Where and how to get information is highly dependent on the specific problem that is being looked at. However, there are some general comments that can be made concerning the basic methods of gathering information. These methods include review of literature, interviewing, sampling, question-naires, observation, forecasting, modeling, and testing. While these techniques are discussed separately, there is much about them that overlaps when using these approaches in actual situations.

Review of Literature

Very few systems problems are totally unique. Often someone else has experienced a similar situation and has written about the results in various periodicals, textbooks, company reports, studies, etc. If this is true of your problem situation, it could be well worth your time to work from these previous results. This will save much effort and money since you will not have to "reinvent the wheel."

However, it isn't always easy to determine if in fact this informa-tion is available and how to go about getting it. Generally speaking, your best friend in this situation is a good reference librarian. These specialists can be found within your organization, at major city

Table 9-2   Review of literature.

What?	Ways to do it
To look at appropriate academic journals, magazines, newspapers, textbooks, government publications , company reports, consultant studies, etc.	Check abstracts, indexes, reference works, library card catalogs, trade associations

Why?	Problems with method
Trying to take advantage of what other people have already done and written about	Establishing similarity of your situation to those reported
	Search is costly in time and energy
"Don't have to reinvent the wheel"	Difficult to get actual reports
	Cost of publications themselves
	Information potentially outdated due to time lag in publication

libraries, and at essentially all college and university libraries. Additionally, depending on the subject matter, experts in the particular field under study generally know of works which could be helpful.

While the advantages are many, the costs associated with a literature review can be expensive. One needs to establish the degree of similarity of the problem situation to those reported. Further, this search procedure is rarely straightforward and generally winds up consuming much personal time and energy. Sometimes appropriate reports can be identified but one can't get hold of them because they involve proprietary information, time delays in receiving mail, cost of publications, etc.

Interviewing

Talking with people close to the situation can be one of the most useful ways of gathering information of various kinds. In a formal sense this can be called interviewing. Generally, there are three main reasons for conducting interviews: first, to gain general background information concerning the problem area; second, to gain specific information resulting in facts and/or best guesses/estimates from knowledgeable people; and thirdly, to develop insight into the competence, feelings, opinions, biases, and/or pressures on individuals or groups.

The type of problem we are trying to solve directly influences whom we want to interview. Basically, we need to look at the individual's background to establish the degree of expertise or experience in the problem area at hand. Further, it will be helpful to get recommendations from people on who will actually communicate with you about what they know or feel.

There are several points which must be made explicit about interviewing as a source of information. Interviewing is a reactive technique. That is, a person will respond differently to the same questions depending on who is asking the questions and how they are asked. Therefore the interviewer needs to adjust his interviewing techniques depending on who he is talking to (executive, worker, staff), who he is (high-level line, middle-level staff, outsider) and what the subject matter is about.

The confidence of the person being interviewed must be earned, not presumed. While the analyst may have determined who can best supply him with information, why should the person being interviewed really tell him what he knows?

Talk is cheap. People generally have poor memories of what has transpired; they tend to view things from a personal perspective, and like to exaggerate their own influence or self worth. This does not mean the analyst cannot learn anything. What it implies is that ideas,

Table 9-3 Interviewing.

What?	Ways to do it
Questioning people with expertise or experience in a particular area	Look at people's background which match problem area
	Consider sources such as consultants, professors, workers, management, competitors, vendors, customers

Why?	Problems with method
To gain from other people's knowledge	Can be time-consuming to find appropriate person
To gain insight into a person's beliefs, opinions, feelings	Can be very difficult getting person to meet with you
"Picking other people's brains"	People don't always communicate what they really feel or believe

opinions, etc. must be cross-checked before they can be assumed to be valid.

In addition to these points, the interviewing process can be very time-consuming. It may take time to determine who the appropriate person is and then work to get the person to meet with you. Further, this meeting doesn't always result in you learning what you wanted to gain from seeing this individual.

Sampling

Many times the analyst will find the technique of sampling useful. The primary purpose of sampling is to get an indication of what a particular situation is without having to gather all the relevant data.

For example, in quality control work, it is desirable to know what percentage of a product line falls below prescribed control standards. One way to determine this information is to check all finished products. But this is generally far too expensive to do. However, by using appropriate sampling techniques, one can get essentially the same insight at a fraction of the dollar cost.

Sampling is also very helpful in assessing customer reaction to new products. Large organizations use the concept of test markets to sample the viability of new products. Sampling can take many forms, to in-

Table 9-4 Sampling.

What?	Ways to do it
Establishing the appropriate sub-population of people, records, or products to examine	Checking historical data in computer-data bases
	Surveying company personnel files
	Reviewing quality control records

Why?	Problems with method
To get a reliable indication on what some particular situation really is without the cost of having to check out all cases or aspects	Results highly dependent on good sample design
	Have to be aware of psychological aspects
"Insight, at a fraction of the cost"	Reliability dependent on number of cases sampled
	Basic information can be confidential

clude analysis of manual and computerized records, surveying people's reactions to certain situations, and using questionnaires to determine the attitudes and feelings of individuals.

The analyst needs to be aware that the results generated are highly dependent on a good sample design. In addition to the sample being truly representative, there needs to be a check on the number of cases necessary for the results to be reliable. To be useful, the sample taken must be representative of the total situation. This can be achieved through the use of random sampling techniques—that is, by insuring that each aspect of the population has an equal chance of being selected. One need then sample only a small percentage of the population to be confident of the results. This has been demonstrated in the TV coverage of most of the recent elections where prediction of the eventual winners has shown a high degree of accuracy when in many cases only 1, 5, or 10 percent of the districts have reported. The same is true of many of the pre-election polls of voter's preference.

Where the surveying of records—especially personnel files—is required, the information obtained can be considered confidential. This information is, therefore, generally much harder to gather and when compiled it must be handled in a way that respects the basic rights of the individuals from whom it was developed.

Questionnaires

Often the analyst will need to know what a large number of people think about particular issues, conditions, or situations. One of the most efficient ways to get this information is through the use of a questionnaire. Basically, a questionnaire asks people to respond to various written questions which are presented in a formalized document.

Questionnaires can either be highly structured in the sense of limiting the responses to questions to predefined entries or they can be much more flexible by allowing open-ended responses. Which is best depends on what the analyst is trying to accomplish and also on the perspectives of the individuals being questioned.

Questionnaires can be very effective in surveying people's interests, attitudes, feelings, and beliefs. However, the accuracy of the results in highly dependent on how well the questionnaire was designed and interpreted.

To develop questions which people will interpret in a consistent manner and in the way the analyst intended is very difficult. One of the best ways to insure the desired results is to use a two-stage phase.

Table 9-5 Questionnaires.

What?	Ways to do it
Asking people to respond to various written questions in a formalized document	Highly structured questionnaire asking person's agreement/disagreement to specific statements
	Open-ended survey asking person's comments on various issues

Why?	Problems with method
To learn people's interests, attitudes, feelings, or beliefs toward particular issues, conditions, or situations	Accuracy of results highly dependent on how well the questionnaire was designed
Generally a very efficient way to get the responses of a large number of people to a consistently phrased set of questions	People answering questions as they think they should, rather than as they really feel
	Time-consuming to develop, administer, and interpret surveys
"If you want to know what people think, ask them"	

First, send out the prototype questionnaire to a representative sample of people to get their responses. Then go talk with them and see what difficulties they had interpreting and responding to the various questions. Also, ask them to suggest other points which should have been asked, etc. From these results, the prototype can be revised and the final questionnaire sent with expectations of a higher degree of accuracy. However, it is time-consuming to develop questionnaires and test their effectiveness.

When it comes to interpreting the results of the questionnaire and reaching various conclusions, the analyst needs to be aware that the questionnaire is a form of sampling. How many and which people to contact will affect the accuracy of the results. Further, the analyst has to consider the psychological aspects of surveys. Any penalties or rewards that will implicitly or explicitly result from answering the questions can sway individuals to give responses that fail to reflect how they really feel.

Observation

Sometimes the best way for an analyst to get information about how a system works, or what level the employee morale is at, or the living conditions of people on welfare, etc., is to go observe it for himself.

By personally observing certain situations, the analyst can gain insights which can't be acquired in any other way. Restricting oneself to the analysis and reporting of other people about a particular situation, limits us to what can be described through written or pictorial means. Further, we will only see what the author of those reports perceived to be important. Especially in the beginning of a study, it is helpful to immerse oneself in the problem area. This process yields much background information that will be very useful throughout the study.

A very important limitation to the observation method is what is known as the "Hawthorne effect." When people are aware that they are being watched, they will tend to act differently than they usually do (i.e., on their best behavior, so to speak). If the analyst is not aware of this, he can draw very erroneous conclusions.

Observation is also a very time-consuming process and moreover the value of the results depends greatly on the perspective of the observer. Being human, many people see only what they want to see. Further, the analyst, even though he is trying to be openminded and unbiased, works from a personal perspective.

What is desired is a set of unobstrusive measures which reflect the actual situation and are independent of the observer. For example, in determining which museum exhibits are most valued by visitors, one

Table 9-6 Observation.

What?	Ways to do it
To observe a situation and draw conclusions to the underlying reality	Personally sit and observe a situation
	Look at film or listen to tape-recording of a situation
	Review unobtrusive measures of a situation.

Why?	Problems with method
A good observation will show how something really works, as opposed to how people say or think it works	Highly dependent on the perspective of the observer
	Hawthorne effect—situation is changed because the observer is there
"Actions speak louder than words"	Can be considered a very limited sample
	Is very time-consuming

could conduct a survey and ask people as they left the museum. You would tend to get answers which sounded socially acceptable. A much more reliable method would be to photograph the people where they actually congregated. If one had a longer time period, he could measure the wear and tear of the floors around each exhibit (Webb et al.).

The analyst, in trying to understand the structural relationships of people in an organization, could first study the organizational chart. However, to get real insight he then would need to go observe an actual committee meeting to see the real pecking order. Another example of how "actions speak louder than words" would be for the analyst to interview the company president or refer to the policy manual to establish what are the highest and most important priorities for this organization. This can and should be compared with the actual budget allocation for the present year to see what the system priorities really are.

Forecasting

Since the primary objective of a system study is to increase the effectiveness of the system during the planning horizon, forecasting has an important part to play. There is a variety of methods for generating estimates of what the expected performance of particular aspects of a system will be during a given future time period. These techniques include extrapolation, expert opinions, and assumptions.

The most commonly used method is extrapolation and its various offshoots. Extrapolation or trend analysis essentially assumes that whatever has been happening will continue to happen in the same way in the future. For example, if sales for a company have been growing at 10 percent per year over the last five years, we would estimate that sales growth would be 10 percent next year. The pro-forma approach in accounting uses this concept in developing future balance statements. If the cost of goods sold has averaged 70 percent of the sales price, we expect that to continue during the next reporting period. Extrapolation is a black box approach which doesn't require or make use of any understanding of the underlying process.

A second approach is to ask experts who have studied the underlying variables and their interrelationships, what they feel will take place in the future with the system. In this approach one is trying to go beyond the outward trend analysis to understand the innards of the black box. With very complex systems and/or when looking out over a long future time period, it is sometimes helpful to query the opinions

Table 9-7   Forecasting.

What?	*Ways to do it*
To generate an estimate of what some aspect of a system will be like in a future time period	Observe what the situation is presently and assume it will continue in the future
	Ask experts for their opinions based on knowledge of the underlying variables and conditions
	Assume a plausible situation

Why?	*Problems with method*
The primary emphasis in a system study is to increase the effectiveness of a system during the planning horizon	There are no facts with regard to the future, just supposition, conjecture, and guesses
"Future isn't always like the present"	To gain a high confidence level is very costly
	Many key aspects are contained in the environment which is outside the decision-maker's control

of several experts. The Delphi method does this in a somewhat formal manner, utilizing a feedback process among the experts. Finally, the last approach is to just basically assume that certain key variables will take on particular values and see what impact that would have on the system. This method is incorporated in the concept of the scenario.

Each of the approaches has serious limitations in predicting just what will take place. There are no facts with regard to the future, since only time will tell exactly what will happen. Unfortunately, that is too late to help in decision making with regard to systems problems. Therefore, the analyst has no choice in trying to forecast what will take place in the future. His choice is whether to make this uncertainty explicit or leave it implicit.

Further, to gain a high level of confidence in the forecast—even where this is possible—is very costly. The analyst needs to recognize that the best that can be hoped for is calculated supposition, conjecture, or guesses regarding the future results and he needs to act accordingly.

Modeling

For those systems problems which are highly complex, it may be a definite advantage for the analyst to build a model of the system. The model would allow the analyst to better understand the complex interrelationships of the system and through the model the analyst could generate specifically tailored information that he could not retrieve from any place else.

There are many different kinds and types of models, but basically they are all abstract representatives of actual systems or subsystems. In the case of a relatively simple model such as a flowchart, there is a very close relationship between what is depicted and what is the working reality. As one moves to organizational charts, however, the model may reflect formal relationships quite accurately but give only a poor idea of actual power relationships.

Mathematical models in the form of equations and computer-based techniques like simulation, linear programming, queuing theory, etc., are very useful in gaining insights into the underlying relationships of systems. However, their use requires a high degree of abstraction from the complexity of the actual situation. This can be either helpful or misleading depending on how representative the model is of reality. One should never assume the correspondence is one to one!

The primary advantages gained by the use of modeling are they aid the analyst in understanding the underlying system relationships, and they permit the user to analyze and experiment with complex situations.

Table 9-8 Modeling.

What?	Ways to do it
An abstract representative of an actual system	Logic
	Flowcharts
	Simulation
	Mathematical relationships

Why?	Problems with methods
A means of better understanding the complex interrelationships of a system and generating expected outcomes to different assumed situations	Generally just allows quantitative considerations
	Can be very expensive and time consuming to build
"Systems are counter-intuitive"	Very difficult to test the validity of the model to the real system
	Inferences from model, highly dependent on the ability of the user

The major limitations to the use of models are they generally just allow the quantitative dimensions of a system to be considered, they are generally very expensive and time-consuming to build and finally, models are an abstract of reality and therefore the information derived must be judged in terms of the real world.

Testing

An analyst has the continual need to verify that a proposal, product, concept, or course of action works as it has been designed to do. Unfortunately, because most decisions on future courses of action have to be made prior to the actual building of the systems design, the analyst has to rely on the methods of forecasting, modeling, etc., which are abstractions of the final situation.

But in those cases where the analyst is also building the design, or where designs have already been built or prepackaged and/or the analyst is responsible as a user for acceptance, there is a need to use the method of testing. In all these cases, the analyst has to get a clear

Table 9-9 Testing.

What?	Ways to do it
To test an alternative design under varying conditions	Experiment with prototypes
	Controlled tests with final product
	Field tests with final product

Why?	Problems with method
To verify the performance of the alternative under certain conditions	Very difficult to get truly controlled conditions
"See how the system really works"	Tests are not always representative of real situation
	Can be very costly and time-consuming

understanding of what the required systems performance should be. Then he must develop methods to see if the performance requirements will be met with the design under consideration. To accomplish these goals, the analyst has to design the appropriate tests. Further, the analyst must then insure a controlled environment to measure the true results of a system as opposed to transient fluctuations.

Effective testing can often be accomplished on submodules instead of the entire systems design. This decreases the complexity and thus the cost of the testing procedure. However, testing all the submodules is not equivalent to checking out the total design. Where possible, it is best to take a top-down approach, checking out the total design and then selective modules.

When testing, it is preferable to have controlled conditions where a limited number of elements are allowed to change. For example, when parallel-testing a computer system, use people who know the computer intimately and others who are knowledgeable about the manual method. Otherwise, if there are discrepancies, you don't know if the problem is the design or the inexperience of the people, etc. It is also desirable to have controlled conditions to permit entering known errors or bugs in a final systems design to see what happens. This helps establish the tolerance of the design and identify what recovery schemes are built-in.

However, the most used tests are those under field (i.e., actual) conditions. Here the product or systems design is put into operation and run by users. Testing the performance after reasonable learning periods can be quite revealing. However, many of these tests are very costly and time-consuming to carry out. Checking out other companies which are similar to yours who have been using the same systems design, can yield accurate information at much less cost and time.

CONCLUSION

The good analyst learns a wide variety of methods for use in gathering information during a system study. But to be effective, one also needs to know when the various methods are appropriate and in turn needs to be aware of the problems associated with each approach.

Class Exercise 9-1

SEARCHING FOR A NEW CAR

Assume you were in the market to purchase a new car.

1. Describe how you could use the adaptive search procedure to finally determine a feasible set of cars to evaluate in depth. That is, what would you do first, then next, etc. When would you stop looking?
2. How would you go about this search procedure if you had four hours to make a decision? If you had four days?
3. How would your procedure differ if this was a car for your personal use versus if you were say the purchasing agent for Avis Rent-a-Car and were making a purchase of 5000 cars?
4. What type of information could you gather on a new car using each of the basic methods of information-gathering as a purchasing agent for Avis:

   (a) review of literature

   (b) interviewing

   (c) sampling

   (d) questionnaire

   (e) observation

   (f) forecasting

   (g) modeling

   (h) testing

Class Exercise 9-2

ESTIMATING FUTURE PERFORMANCE

You have decided to coach a youth soccer team next year. There will be a draft held two months prior to the start of the season, where each coach selects his team from a pool of ten- and eleven-year-old boys. In talking with people who have coached before, you have gathered the following methods for prediction of future soccer performance of the kids.

(a) Coach A says he uses the physical characteristics of the boys. That is, the age, height, and weight.

(b) Coach B goes by the coaches' ratings of the previous year. That is, each coach rates each individual of his soccer team as to overall performance. The ratings are from 1 to 5 with 3 being average and 5 being super.

(c) Coach C feels that because a boy's performance changes so much over a summer because of growth and maturity, the only useful rating system is for the boys to have a workout session. All coaches could be present and then they would rate the players themselves. The test would be

(a) a 50-yard dash,

(b) maneuvering the ball around a series of cones,

(c) making ten kicks at the goal from various lengths and angles, and

(d) playing goalie and trying to save 5 kicks.

(d) Coach D feels the only reliable test is to have the boys play the various positions in a game situation. So he holds a series of practice games in which players play front, middle, back, and goalie positions.

1. For each of the coaches' plans, classify them according to the basic methods of information gathering, i.e., review of literature, interviewing, etc.

2. Discuss how good you think each method is in predicting future performance. What is the cost associated with each method?

3. The new commissioner of the league has asked which method or methods would you recommend the soccer league adopt. What is your rationale?

4. Compare the methods of prediction used in the soccer example with those commonly used in predicting future computer performance. That is, benchmarks, demonstrations, kernels, and references. What is the relationship?

Class Exercise 9-3

DETECTING SOURCES OF OVERPAYMENTS

Assume you have just joined a state welfare agency as a systems analyst, and were told that overpayments were a major problem area. You have been given the task of establishing the relative importance of overpayments to welfare recipients in terms of computer errors, inaccuracies in claim forms submitted, and fraudulent practices, etc.

1. How would you go about this task?
2. How could you use the method of sampling?
3. How would observation be used to establish fraudulent practices? Does this conflict with basic privacy considerations?
4. How would you go about interviewing people who keypunch the data into the computer? Assume this turns out to be a major source of the errors?

Chapter Ten

What Information to Gather

In this chapter, we will discuss the characteristics of system studies and show how they differ from the much more commonly performed research studies. The major elements within a study will be covered and discussion will center on how these aspects are treated in the four major reports of a system study. Lastly, the need for a problem-solving group and its composition and influence throughout a study will be presented.

SYSTEMS STUDY PURPOSE

It is extremely important to emphasize and reiterate that the purpose of doing a systems study is to make a particular system more effective. This can only happen if, in fact, the system is changed in accordance with the results of the study. Thus, the aim is to change the system not just study it!

To help differentiate a system study from the more usual study, let's name some actual system studies that have been completed using the systematic systems approach. Considering individual and family systems: What is the best law school for John Jones to attend? What is Bob Smith's best career? What is the best religious position for Ed Ames? What is the best life style for the Browns?

Studies concerning business and organizational systems have been: What is the best inventory policy for the ABC stores? What is the best medium-size computer for Acme Hardware? What is the best way to market copperware from Iran in southern California? What is the best organizational structure for the Beta-Beta fraternity? What is the best remedial reading program for use in the XYZ School District?

At a more complex systems level, studies have been done on: What is the best way to finance presidential political campaigns in the United States? What is the best energy policy for the United States during 1985-2000? What is the best way to handle the Middle East situation looking at it from the Israeli position?

In addition to noting the very wide scope of topics to which the systems approach can be applied, compare the way in which emphasis differs from the more usual research papers on the same subject matter. For example, What is the situation in the Middle East? What have the abuses been in campaign financing? What is the inventory policy of Acme Hardware?, etc.

While this latter type of inquiry can produce very useful knowledge, it is *just information-gathering* and *not problem-solving*! At most, it would be background information for a systems study. Therefore, a systems study is concerned with solving some problem. A decision is required at the conclusion of the study, on how to best restructure the system for the future.

DOING A SYSTEMS STUDY

As has been initially discussed, the systems study is composed of preliminary, feasibility and evaluation reports which unfold toward the final report containing the recommended solution to the systems problem. We now consider in detail what elements are covered in each report.

Preliminary Report

The purpose of this initial look at the system is to ascertain what the problem really is and to determine what steps should be taken to resolve it. The major elements comprising the preliminary report are shown in Fig. 10-1.

The purpose of these elements are as follows:

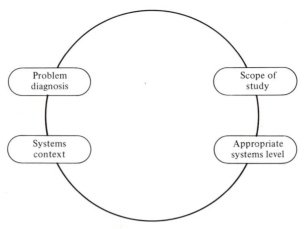

Figure 10-1 Preliminary report.

Problem diagnosis—determine the cause or nature of the problem or situation.

Systems context—ascertain what the most relevant characteristics are of the system that encompasses the problem area.

Appropriate systems level—determine the system boundaries as a function of the composition and authority of the problem-solving group.

Scope of study—decide to what breadth and depth the system study should go.

Feasibility Report

The feasibility report is based on the preliminary report in that it contains the problem diagnosis, systems context, appropriate systems level, and scope of the study (see Fig. 10-2). Additionally, the following new aspects of a systems study are considered:

Constraints—determine what conditions or restrictions any potential system solution must meet for it to be considered workable (viable).

Criteria—conclude which aspects of the overall system objectives should be used to evaluate the worthiness of alternative designs.

System designs—learn which particular selections and arrangements of system components will solve the stated problem.

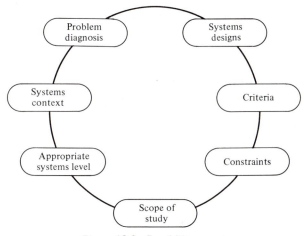

Figure 10-2 Feasibility report.

The major purpose of the feasibility report is to use the additional elements of constraints, criteria, and system designs to establish if there are any workable solutions to the stated system problem.

Evaluation Report

The purpose of the evaluation report is to demonstrate the overall desirability of the alternative system designs that have been advanced. This is accomplished through reconsidering the previously established study elements of problem diagnosis, systems context, appropriate systems level, scope of study, constraints, criteria and systems designs. The new aspects covered in this report are the desired performance, expected performance, and alternative rankings. See Fig. 10-3.

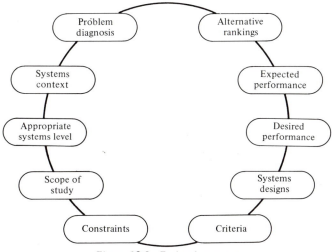

Figure 10-3 Evaluation report.

These elements are defined as follows:

Desired performance—establish the relative desirability of various levels of systems performance from an overall perspective.

Expected performance—estimate the future level of performance for each system design.

Alternative rankings—conclude on the set of system designs to be evaluated, what their relative rank is according to overall worth.

Final Report

The purpose of this report is to show the best way that has been determined for handling the problem and in so doing, increase the overall effectiveness of the system. This is accomplished by a final reconsidering of study elements of problem diagnosis, systems context, appropriate systems level, scope of the study, constraints, criteria, systems designs, desired performance, expected performance, and alternative rankings.

The new aspects of the final report concern how to implement the problem solution and establish the new system goals. This is accomplished by selecting the best alternative system to implement, planning systems change, and establishing systems goals. These elements are defined as follows:

Select alternative—resolve which system solution is best by considering the overall performance of each alternative design and the explicit tradeoffs and risks involved with each choice.

Plan system change—learn how to implement the problem solution to insure the desired effect.

System goals—establish, once the solution is implemented, what the expected level of system performance should be.

The concepts of select alternative, plan system change, and system goals will be discussed in detail in chapters 13, 14 and 16.

The total system study process is shown in Fig. 10-4. This summary chart shows how the preliminary, feasibility, evaluation, and final reports are all interrelated. Second, the cycling nature of a system study is shown by the circling arrows which also depict a continually expanding area of knowledge about the problem and its solutions.

The extensions to the small circles, which represent major elements in a system study, are used to reinforce the iterative nature of a study. Be sure to note, however, that this does not signify duplication! For example, the initial problem diagnosis may have to be revised several times as new information becomes available in the later stages of

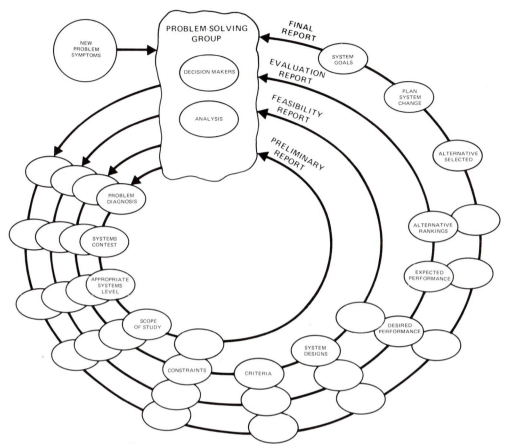

Figure 10-4 Iterative approach to a system study.

the systems study. The expected performance of future alternative designs during the evaluation report may suggest that the appropriate systems level be substantially increased, or modified in other ways.

Therefore, recycling through the various stages and elements does not mean going over the same ground. Additional insight should bring about better problem definition, more encompassing criteria, more accurate levels of expected performance of the alternatives, etc. On the other hand, if an analyst is very well satisfied with a particular stage of the study, those particular elements can be covered very lightly on succeeding cycles.

SYSTEMS STUDY REPORTING FORMATS

We have covered the overall philosophy of the adaptive search procedure as it relates to doing a systems study. In turn, we have covered the unfolding nature of the systems study through the preliminary, feasibility, evaluation, and final reports. In the previous section the purpose and major elements of each of these reports were discussed.

Now the specific contents of each report will be covered in more detail and the appropriate format will be shown.

Preliminary Report

The basic purpose of the preliminary report is to establish what the systems problem appears to be and what level of effort should be applied to solving this problem. This report should be done relatively quickly and emphasis should be placed on getting back to management with the results, however tentative they are. The time span should be approximately 5 percent of the total study time.

A suggested format of the preliminary report appears in Table 10-1. The problem diagnosis/(section I) contains short statements regarding the symptoms, cause, and identification of the problem. Potential solutions are rendered, however tentative, at this point to help clarify the situation for the CDM and others.

The systems context referred to in section II is a listing of the key variables or components in the system and a spelling out of their present state. This can, but need not, be shown through a systems context diagram (see Fig. 2-1).

Section III defines the system and indicates who is the controlling agent (CDM). This could be one person or an appropriate committee. Further, an estimate of the subject complexity and the potential for making a difference with this problem is given.

The last section is the scope of the study. This concerns establishing the problem importance and assuming a relevant planning horizon. The criteria that must be established are when a decision must be made and the amount of resources that will be allocated for the study.

Feasibility Report

The purpose of the feasibility report is to further refine the concept of what the problem is and what effort should be put forth in

Table 10-1 Preliminary report.

I.	*Problem Diagnosis*	*III.*	*Appropriate System Level*
	Symptoms		CDM
	Cause		System
	Problem identification		Subject complexity
	Potential solutions	*IV.*	*Scope of Study*
II.	*System Context*		Problem importance
	System context diagram		Planning horizon
	Present state of key variables		Time to decision
			Study resources

Table 10-2 Feasibility report.

I.	*Problem Diagnosis*
II.	*System Context*
III.	*Appropriate System Level*
IV.	*Scope of Study*
V.	*Constraints (major)*
	External
	Internal
VI.	*System Objectives*
	Technical criteria
	Economic criteria
	Psychological criteria
	Political criteria
	MAN
VII.	*System Design Alternatives*
	Key components
	Feasibility

deriving a solution. Further, this report is to establish if in fact there are workable solutions to the problem.

The sections for this report are shown in Table 10-2. The first four sections cover the same items that were discussed in the preliminary report. However, this time the answers should reflect more intensive analysis of the problem for about 20 percent of the systems study time allocated.

The new elements introduced in the feasibility report are delineating the major constraints and specifying whether they are internal or external. While this latter step is not required, it does help the analyst in the early stages of a study to note what restrictions can be changed if really necessary. Section VI is concerned with system objectives, more specifically the relevant criteria. As an aid to understanding the situation, it is suggested that the analyst specify the criteria in terms of their technical, economic, psychological and political aspects. If desired, this consideration can be dropped in the latter stages of a systems study.

A set of feasible alternative solutions is given in section VII. Each of these systems designs will solve the stated problem and would be checked to see that they satisfy the constraints specified in section V. While it is desirable to get to the final set of alternatives that will be evaluated in depth, the alternatives specified in this report are understandably subject to change based upon further knowledge gained in the later parts of the systems study.

Table 10-3 Evaluation report.

I.	*Problem Identification*
II.	*Systems Context*
III.	*Appropriate Systems Level*
IV.	*Scope of Study*
V.	*Major Constraints*
VI.	*Systems Objectives*
VII.	*Feasible Alternatives*
	Final set
VIII.	*Desired Performance*
	Criteria
	Performance measures
IX.	*Expected Performance*
	Outcome
	Information support
X.	*Alternative Rankings*
	Total value
	Discounted value

Evaluation Report

The first seven sections of the evaluation report Table 10-3 contain refinements of the elements included in the preliminary and/or feasibility reports. This report covers work that has been accomplished through approximately 40 percent of the allocated study time.

A new item covered in section VIII is the desired performance of the system. This is shown through the redefining of the major criteria and by specifying the measures of performance. In chapter 12, this will be covered by the systems utility function. Expected performance is covered in section IX. This includes showing the projected systems outcomes for the alternatives specified in section VII of this report and further giving the evidence to support conclusions of the analysts. The systems simulation chart can be used here.

Section X covers the initial look at the ranking of the alternatives according to total value and discounted value. This is a function of comparing each alternative as to its expected performance in relationship to the desired systems performance as shown in sections VIII and IX. The evaluation matrix as explained in chapter 12 is a way of doing these comparisons.

Table 10-4 Final report.

I. *Management Summary*

II. *Problem Identification*
   Symptoms, cause, problem, scope of study

III. *Systems Context*
   CDM, system diagram, scenario, systems objectives

IV. *Feasible System Designs*
   Major constraints
   Final alternatives

V. *Desired System Performance*
   Criteria
   Preference chart
   Systems utility function

VI. *Expected Systems Performance*
   Outcomes
   Confidence
   Information support
   Systems simulation chart

VII. *Recommendation of Best Alternative*
   Alternative rankings
   Decision rationale
   Tradeoffs

VIII. *Planned System Change*
   A Pressure
   B Relative advantage
   C Goal congruence
   D Amount of behavioral change

IX. *Systems Goals*
   A Measure of effectiveness
   B Systems goals

X. *Information Sources*

Handwritten margin notes:

1 PROBLEM IMPORTANCE
2 PLANNING HORIZE
3 TIME TO DECISION
4 STUDY RESOURCES
+ MANG.

D POTENTIAL SOLUTION
E SYSTEM CONTEXT
  1 STATE OF KEY VAR
F APPROPRIAT SYSTEM LEVEL
  1 CDM
  2 SYSTEM
  3 SYSTEM COMPLEXITY

POSITIVE MODIFIERS
BONUS + INCENTIVE
PT COMMITMENT

Key Variables

AND OBJECTIVES

* III MAJOR CONSTRAINTS
A EXTERNAL Con
B INTERNAL CON
C. SYSTEM OBJECTIVES
  1 TECHNICAL CRITERIA
  2 ECONOMIC ''
  3 PSYCHOLOGICAL ''
  4 POLITICAL
  5 MANAGEMENT

Final Report

In Table 10-4, the content of the final report is given. Sections II through VII combine all the elements covered in the preliminary, feasibility, and evaluation reports. Some of the sections have different headings to reflect how different subelements are now combined. For example, section II on problem identification now covers symptoms, cause, problem statement, and scope of the study. The potential solutions, which have gone through considerable change and refinement,

167

are discussed in detail now in section IV, feasible system designs.

New to this report is section I, management summary. This is a one-page overview of the total study and a summarization for management of the results and recommendations.

Section VII now contains not just the alternative rankings, but also the decision rationale behind the selection of the best solution and an explicit statement of the required tradeoffs. Section VIII, planned systems change, covers the readiness of the system for change. The impact of the recommended solution in terms of the pressure, relative advantage, goal congruence, and amount of behavioral change of the system. This will be discussed in chapter 14. The establishing of system goals is covered in chapter 16. These ideas for the recommended solution are specified in section IX.

The last section is a bibliography. But since we are not limiting ourselves in a system study to just "books," we will call this section X, information sources. It identifies all major sources of data used in the study, comprises a reference list so other people can follow up on these thoughts, and provides an overall confidence indication for each source.

SCOPE OF THE STUDY

Those elements which comprise the various preliminary, feasibility, evaluation, and final reports are explained and defined throughout the text. What has not been explicitly covered any place else is the scope of the study.

Once the central purpose of the study has been established and the level of the CDM assumed, it becomes necessary to determine the scope of the study. While it would be nice to be able to study each problem to its fullest and determine the "best" solution, the reality is much different. Time is always short, resources are limited, the "best" solutions can't always be determined, etc. Therefore, the analyst and CDM need to establish what kind of effort is warranted in deriving solutions to the problem at hand.

The scope of the study concerns the range and depth of the inquiry into the problem and its solution. This is established by considering the problem importance, planning horizon, time to decision, and study resources.

Problem Importance

The concept of problem importance was covered in detail in chapter 4 so we will summarize the main points here. In general, strategic problems are considered more important than tactical problems. Operational problems are considered least important in terms of the relative impact each type of problem has on the system over an extended period of time. Assessing the consequences of doing nothing or the risk of making a bad decision, also influences the determination of the problem importance.

Planning Horizon

This factor was considered in depth in chapter 3 and is therefore summarized here. The assumption as to what future time period the analyst expects the implemented solution to be effective (i.e., a day, year, five years, or more) is called the planning horizon. In general, the longer the planning horizon, the more factors that can affect the value of the system solution.

For example, consider the selection of the best job today for a particular student just graduating from college compared with determining the best job for him over the next twenty years. The considerations that need to be made in these two cases are highly distinct. The kind and amount of data needed and the accuracy of the information available are substantially different.

The range and scope of the study should reflect what time period is being projected and how much effect changing conditions can have on the proposed solutions.

> **Planning horizon** is the assumed time period for which any system solution must be effective in order to be acceptable.

Time to Decision

A major factor in deciding to what depth a study should proceed is the consideration of how long one has to study the problem. This, in turn, depends on when a decision must be made to implement the system solution. Many times when a decision must be made, it is determined either by higher authority or by when the system will become infeasible. In any case, an analyst will attack any given problem in different depth depending upon whether he has a day, a week, or a year in which to attain a system solution.

> **Time to decision** is the latest point in time in which the decision can be delayed without adversely affecting the system being studied.

An example of time to decision would be when you applied for college, and the college stated that the last day for acceptance was say August 28, 1980. In other words, you could wait no longer than this time before making a decision on going to college starting in Fall 1980.

We are often advised that to be a good manager we will need to make decisions quickly. But it would seem to be wiser to be able to differentiate between when it is smart to make a decision quickly about a particular systems problem, and when it pays to delay the decision to a better time.

As an example of this point, consider the plight of the captain of a large jet airliner who, right after takeoff, has problems getting one of the wheels to retract fully. In this emergency situation, where some 200 passengers are aboard, the pilot could make an immediate decision to solve his problem by turning the plane around and attempting a landing at the airport.

Though it is an emergency situation, why reach a decision so quickly? When must a decision really be made? Since the aircraft has 6 hours of fuel left, the decision can be delayed 5¾ hours. By understanding this, the pilot can circle the airport, get his thoughts composed, and have the runway prepared for this type of landing; or he can have the airport get in contact with an expert on landing gears, etc.

Thus, the point of the time to decision is not necessarily to delay decisions as long as possible, but rather to see what time is really available so that the analysis of the problem can be adjusted to when a decision should be made.

Study Resources

The amount of manpower and money available to study a problem also affects the depth and range of a study. Ideally, the number of people, their talent mix, and the money for travel, information gathering, computer time, etc., are commensurate with the type and importance of the problem to be studied. However, this problem is just one of many that the organization is attempting to cope with. Because there always seems to be more problems than resources to attack them fully, any particular system study gets less than what is required to generate an ideal solution. Since this is true, the analyst needs to adjust the range and depth of his study to best utilize the limited amount of resources given him.

In the extreme, if no money is available for traveling to a plant site to observe the problem area in depth, then information must be gathered from secondary information sources like formal reports and telephone conversations with key personnel. While secondary sources may be less accurate than first-hand observations, this may be the best that can be accomplished with a smaller-than-needed travel budget. In turn, the validity of the system study will generally then be less than it could have been with more resources.

Study resources are the amount of time, manpower, dollars, etc., which are allocated for the solving of a problem.

The amount of resources that can be effectively used in a study is a function of the impact on the system effectiveness that a study could bring. Ideally,

Amount of resources to expend $\approx$ Impact of study on system
effectiveness

The impact of studying the system is not only a function of the importance of the study, but also must consider the amount of useful knowledge that can be gained. For instance, strategic problems are considered the most important and, therefore, if importance were the only criterion, all resources should be applied to this type of study. However, in the case of a problem of extreme systems complexity, we can't expect to really learn anything of practical value. So even though the problem is vitally important, if we can't hope to have an impact on it, why study it? Our limited resources can be better applied someplace else. As a rule, then,

Amount of resources to expend $\approx$ Level of decision $\bullet$ Amount of
knowledge

For example, take an operational problem which we can fully understand in all its aspects. Although the problem is low-level since we can have a major impact on the outcome, it may turn out to be a better use of resources than to tackle a strategic problem which we have little hope of understanding. Thus,

Operational $\bullet$ Certainty $>$ Strategic $\bullet$ Total ignorance $\approx$ Amount of
resources to expend

Therefore, the amount of resources that should be expended on a system study is established by determining the level of decision that will be required and the amount of impact we can expect to have by studying the systems problem.

Distribution of Study Effort: Regardless of the amount of time needed to do the study, the analyst always has to adjust his effort to the time available (or granted). As an initial look at this concept, consider the chart (Table 10-5) on distribution of effort required in doing a system study.

The time an analyst should spend on each part of the study breaks

Table 10-5 System study distribution of effort.

	Percentage of Time	Pages	Effort
Preliminary report	5%	1	2 days
Feasibility report	15%	2-3	6 days
Evaluation report	30%	4-6	12 days
Final report	50%	15-25	20 days
			40 days

down approximately into 5% for the preliminary report, 15% for the feasibility report and 30% for the evaluation report. One half of the time should be allocated to the completion of the overall final report. Similarly, if the overall system study writeup would require 15-25 pages (some studies require 100 pages, others thousands of pages depending on the importance of the project) the preliminary report should only be about one page. Furthermore, if the analyst was only given forty days to do a system study, no more than two days should be spent on completing a preliminary report.

These figures are not meant to be hard and fast rules, but rather are given to help foster the idea of a top-down approach which progresses toward the best problem solution. Hopefully, this timing will encourage the analyst to keep the big picture in mind in the early stages of the study and not get bogged down in the details.

PROBLEM-SOLVING GROUP

Scant attention has been paid so far to what the purpose of the problem-solving group (PSG) is, what its composition is, and how it functions throughout the system study (see Fig. 10-4).

Composition

The main purpose of the PSG is to determine the best way to solve the systems problem. This group is composed of people who will need to function as decision-makers and analysts. In the case of systems problems involving individuals and small groups, one person might function as both the analyst and the decision-maker. On the other hand, a major problem associated with a large, sophisticated organization, there might be twenty or thirty analysts and ten to fifty decision-makers. Regardless of the number of people involved, the basic functions of the PSG will be the same.

172

For example, consider the case of a large manufacturing organization deciding the best way to establish a computerized information system to aid in the management of the firm. Because of both the technical nature of this subject matter and the fact that it will affect every major functional area of the organization, the PSG would most likely be very large. For decision-makers it would include people like the vice-presidents of marketing, production, finance, administration, etc. Analysts would be required from each functional area. Additionally, analysts with technical expertise in computers and information systems would be needed. Incidentally, all people within the PSG would not have to be members of the organization. For example, consultants with knowledge in specific areas could function as analysts within the PSG for specific periods of time.

The PSG functions continually throughout the system study, but its main decision points come at the conclusion of each of the preliminary, feasibility, evaluation, and final reports. A formal review is made of the progress of the system study. Critical assumptions are reviewed and expected progress for the next phases are considered. In a sense, each phase has a Go/No Go decision to it. That is, the PSG must decide if progress is being made according to its expectation and whether or not additional effort should be expended and, if so, in what specific directions.

Incidentally, the PSG is frequently designated by names like steering committee, guidance group, project management, ad-hoc committee, etc. We will use the term problem-solving group since it more readily denotes the purpose of the group.

Intergroup Influence

As shown in Fig. 10-5, the relative influence of decision-makers as a group, and analysts as another group, all within the PSG should change over the life of the systems study. As the problem is initially

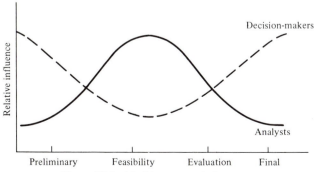

Figure 10-5 Ideal intergroup influence.

established in the preliminary report, the decision-makers should have the most influence on deciding what is studied and to what depth. However, in generating alternative designs and establishing their relative rankings, the analysts should generally be more influential than the decision-makers.

However, in the final report where the selection of the best alternative is decided and also the means of implementing this solution is laid out, the decision-makers should be much more influential than the analysts.

It is worthwhile to compare and contrast this thinking on ideal intergroup influence throughout a system study with that shown in the exercise at the end of this chapter—completed staff work.

CONTINUAL COMMUNICATION IS NECESSARY

Without belaboring it, we cannot overstress the point that the systems approach requires continual communication between the systems analyst and others. In gathering information about the system and how it functions, the analyst needs to communicate with management, staff, and workers. However, most of management and operations research theory and much business practice assumes that there needs to be little communication, for example, between the systems analyst and the CDM after the initial problem assignment. That is, the CDM calls in the analyst, states to him the problem and objectives to be accomplished and then he is told to go solve it. Further need for clarification on the part of the analyst presumably shows lack of understanding, no initiative, or basically no concept of what completed staff work means. While this may sound logical, how effective is this practice?

Under conditions of partial ignorance, how can a manager know what the problem really is without analysis? I have personally seen millions of dollars wasted by analysts studying the wrong problem because they were not allowed, able, or encouraged to get back to the CDM at various stages in a system study.

At the start of a study, how can the CDM state his objectives so concretely, when, in fact, no one really has any knowledge of what is possible or what tradeoffs are involved? To require the CDM to specify in great detail exactly what he wants accomplished would require of the CDM (1) tremendous amounts of time, (2) a great knowledge of the problem area, and (3) a great insight into his own mind of what he really wants or should want to accomplish. This level of specificity would require the CDM to essentially solve the problem, which is unproductive of management's time. It is much more efficient to have the CDM and analyst work as a team where repeated communication is necessary and the level of thought and specificity (detail) increases as more insight into the problem and circumstances are developed.

When working on problems of major significance, one cannot immediately begin to model or even get a handle on the total situation. Therefore, assumptions have to be made in order to decide what variables to include and exclude in the system model, how close to reality the model must be to provide insights, what future situations should the study be geared toward, etc. The list of assumptions and conditions can go on and on. The major point here is who is to make these assumptions? It is not a question of whether or not assumptions should be made. They must be made. But who best has the knowledge of which way to go? Surely it is not the analyst! The analyst may have great expertise in system techniques but probably has little experience in the particular problem area. Why not try to utilize the years of experience of the CDM and the expertise of particular staff people?

In summary, there needs to be much communication between the CDM and systems analyst because (1) the CDM generally cannot state what the problem really is without analysis, (2) the CDM's actual objectives are not really known until alternatives are presented, and (3) many crucial assumptions need to be made in any analysis.

Class Exercise 10-1

COMPLETED STAFF WORK

The memo that appears as Fig. 10-6 was written by an Army officer to his staff, describing what his basic policy was concerning problem solving procedures. The concept of *completed staff work* is a widely followed management principle today in most large organizations in the fields of business, government, and education.

Questions

1. As depicted in the memo, what would the composition of the problem-solving group be?

2. Draw a graph which best typifies the thinking shown in the memo concerning the relative influence of the decision-maker and his analysts throughout a system study.

3. Compare and contrast the graph developed in question 2 with the chart presented in the text (Fig. 10-5).

4. Which procedure do you feel is the more ideal? Give support for your conclusions.

5. Could both procedures be correct under different conditions? What are the basic assumptions of each approach?

6. What is the basic management concept consistent with completed staff work?

The following interesting and instructive paper is being distributed to officers of the Provost Marshal General's office and school:

1. The doctrine of "completed staff work" is a doctrine of this office.

2. "Completed Staff Work" is the study of a problem, and presentation of a solution, by a staff officer, in such form that all that remains to be done on the part of the head of the staff division, or the commander, is to indicate his approval or disapproval of the completed action. The words "completed action" are emphasized because the more difficult the problem is, the more the tendency is to present the problem to the chief in piece-meal fashion. It is your duty as a staff officer to work out the details. You should not consult your chief in the determination of those details, no matter how perplexing they may be. You may and should consult other staff officers. The product, whether it involves the pronouncement of a new policy or effects the established ones, should, when presented to the chief for approval or disapproval, be worked out in finished form.

3. The impulse which often comes to the inexperienced staff officer to ask the chief what to do recurs more often when the problem is difficult. It is accompanied by a feeling of mental frustration. It is so easy to ask the chief what to do, and it appears so easy for him to answer. Resist that impulse. You will succumb to it only if you do not know your job. It is your job to advise the chief what to do, not to ask him what you ought to do. He needs answers, not questions. Your job is to study, write, restudy, and rewrite until you have evolved a single proposed action—the best one of all you have considered. Your chief merely approves or disapproves.

4. Do not worry your chief with long explanations and memoranda. Writing a memorandum to your chief does not constitute completed staff work, but writing a memorandum for your chief to send to someone else does. Your views should be placed before him in finished form so that he can make them his views simply by signing his name. In most instances, completed staff work results in a single document prepared for the signature of the chief, without accompanying comment. If the proper result is reached, the chief will usually recognize it at once. If he wants comment or explanation, he will ask for it.

5. The theory of completed staff work does not preclude a "rough draft" but the rough draft must not be a half-baked idea. It must be complete in every respect, except that it lacks the requisite number of copies and need not be neat. But a rough draft must not be used as an excuse for shifting to the chief the burden for formulating the action.

6. The "completed staff work" theory may result in more work for the staff officer, but it results in more freedom for the chief. This is as it should be. Further, it accomplishes two things:

    (a) The chief is protected from half-baked ideas, voluminous memoranda and immature oral presentations.

    (b) The staff officer who has a real idea to sell is enabled to more readily find a market.

7. When you have finished your "completed staff work," the final test is this:

    If you were the chief, would you be willing to sign the paper you have prepared and to stake your professional reputation on its being right?

    If the answer is in the negative, take it back and work it over, because it is not yet "completed staff work."

Figure 10-6 Completed Staff Work Memorandum.

Table 10-6 Computer acquisition plan.

Phase I	Determine Company Needs
	Company background
	Determining applications computerized priorities
	Present system methods
	Company profile
Phase II	Establishing Computer Vendors
	Computer system requirements
	Disk size requirements questionnaire
	Vendor screening questionnaire
	Final set of vendors
Phase III	Select Best Computer System
	Vendor demo results
	Reference checks
	Evaluation matrix
	Computer vendor selection
Phase IV	Signing Contract
	Detail requirement specifications
	Contract analysis
	Final cost proposal
	Financial arrangement
	Contract negotiation
Phase V	Preparing for Implementation
	Initial software testing
	New workflow procedures
	User education & training
	Site checkout & preparation
	Convert files
	Order forms
Phase VI	Company Implementation
	Hardware delivery — sign off
	Hardware/performance test
	Applications software acceptance test

Class Exercise 10-2

COMPUTER ACQUISITION PLAN

Table 10-6 shows a plan used by a consultant for acquiring a computer system.

1. Do the six phases make sense for this type of acquisition? Are they in the right sequence?

2. How does the acquiring of a computer compare or differ with doing a system study? Is this plan linear or iterative?
3. Put the six phases suggested into the format of preliminary report, feasibility report, evaluation report, and system study.
4. What should the relationship be between the company president, office manager, clerks, and consultant? Who should do what?

Chapter Eleven

How to Evaluate Information

We have emphasized that the systematic systems approach is an information-based methodology for problem-solving. The types of information that are required include knowledge of the systems context, future environmental situation, constraints, criteria, systems performance, and implementation considerations. As we learned in chapter 9, this information can be derived using basic methods like review of the literature, interviewing, sampling, questionnaires, observation, forecasting, modeling, and testing.

In this chapter, we will discuss why this information can't be taken at face value, and how to evaluate its potential accuracy. Further, a means for indicating the degree of confidence an analyst has in regards to a particular situation will be given.

WHY IS THIS NECESSARY?

In coping with system problems, especially of the complexity we have been discussing, the analyst will find himself dealing with many types of information. We generally think of information simply as the facts surrounding a situation. In practice, however, information arises in different shades not just true or false.

We have seen that in the course of the system study we need to develop information which includes facts, opinions, feelings, hopes, and desires. Further, we need to evaluate this information with respect to past, present, and future considerations.

Information regarding historical happenings has a much better chance of being verified as to what, in fact, took place. But even with this straightforward type of analysis, the complete facts can be very hard to come by. Moreover, the thrust of system studies is not verifying the past, but aiming a system toward the future. However, in the future—unlike the past—there are no facts at all. We might be very confident, but never certain, regarding what is going to develop. In the last analysis, the best we can hope for is supposition, conjecture, and educated guessing, with varying degrees of accuracy.

We have said so far that our potential knowledge about systems, even if we had unlimited resources to gather information, is always less than 100% accurate. We can never be completely assured that we have determined the truth or correctness of a situation.

But system studies are not made with unlimited resources. Studying a system is very costly in terms of time, dollars, and manpower. We saw in chapter 10 that the resources allocated for studying a system is highly dependent on the time to decision. As a result, the analyst may have five seconds, thirty minutes, two days, one month or three years to study the situation and determine the best solution.

Our overall point is that when the time to study a situation is less than what is needed, we can expect that there will be much uncertainty surrounding the accuracy of the information gathered.

Complicating even further the potential accuracy of our study, is the possibility the information that is gathered has been purposely falsified. People don't always say what they really believe and feel. The more a person has to gain or lose by what is said, the more likely the information given will be slanted. In the section of this chapter on determining whether information makes sense, examples will be given on how we can be misled and how to recognize this situation and deal with it.

In summary, then, it is necessary to evaluate information for several reasons: uncertainty of information, limited study resources, and purposely deceptive information.

ANALYST'S PROBLEM EXPERTISE

In addition to the amount and type of information that is available concerning a system situation, there is the competency of the analyst to determine the relevant information in the first place, and then secondly to interpret what it means. For our purposes, assume there are three cases. One where the analyst has a high degree of expertise in the subject field, secondly, one where the analyst has no expertise, and the third, where the analyst has limited expertise. In reality, like almost everything else, an analyst's expertise is really a matter of degree as opposed to being an expert/not expert. But these three designations will help highlight the important differences in approach needed by the systems analyst.

Case I: Full Expertise

Consider a situation where the analyst is assumed to be an expert. This could be either because the system under study is one in which he is very familiar through experience, or it is his major field of study, or it is an area which works under the same general principles as the field in which he has expertise. Additionally, it is assumed that the analyst has adequate time to apply his expertise to the subject matter under study.

As shown in Table 11-1, when the analyst has expertise and adequate time, then he can deal with the information itself. From the content of the information and how it was derived and the resultant reliability and validity of the methodology used, the analyst can test the correctness of his conclusions. The source or authority who gave the information is not important.

Case II: No Expertise

The second case involves the analyst who has no expertise in the field. This could be due to lack of background in the subject matter;

Table 11-1 Analyst's problem expertise.

	Case I *Full Expertise*	Case II *No Expertise*	Case III *Limited Expertise*
Situation	Primary field and Adequate time	Outside primary field or Inadequate time	General background and Time to Study
Approach	Evaluate infor- mation itself Forget information source	Accept information Evaluate information source	Evaluate information Evaluate information source

or the analyst can have a good general background, but due to limited time to study the problem he has in effect no expertise. In these situations, the analyst can't rely on his own determination of the validity of the information itself. Rather, he is highly dependent on evaluating the source as a means of validating the information. By using multiple sources to cross-validate information and by assessing the worthiness of each source, the analyst can be quite effective in reaching conclusions about the validity of the information even with no expertise in the field.

Case III: Limited Expertise

The third case is that of the analyst with limited expertise. This situation can come about when the subject matter is not specifically in the main field of the analyst, but his background is sufficient that he can within the allotted time develop the necessary competence. In this situation, the analyst can make many adequate judgments on the worthiness of the information, but he will be even more accurate if he supplements his conclusions with an analysis of the information source.

The purpose of the three cases discussed in Table 11-1 is to show the impact of assuming the analyst is less than fully expert. Further, my own experience indicates that analysts are called upon to solve problems which, in at least 50 to 60 percent of the cases, they have limited or no expertise. In such cases, it makes good sense for the analyst to have a well-rounded education in addition to his technical specialization so that he can, when needed, quickly increase his expertise in a particular problem area. In addition, another way which is discussed below is to be able to effectively evaluate the sources of information.

EVALUATING INFORMATION

In Table 11-1, three different situations were developed in regard to the problem expertise of the analyst. We now focus on the case where the analyst is an expert in the subject area and thus can rely on his own judgment.

In this section, we want to consider how an analyst can evaluate the correctness or accuracy of the information itself. Table 11-2 shows the overall method of approach, which is divided into two phases: establishing what the information is and then determining if the information makes sense.

When gathering information using the basic methods of review of

Table 11-2 Evaluating information.

What is information?	Does information make sense?
Spell out content	How was it derived?
Summarize into a meaningful form	How reliable is methodology used?
	How valid are the results to your situation?

the literature, interviewing, questionnaires, etc., one tends to become overwhelmed with a mass of notes, articles, tapes, etc. At a certain point in the problem-solving cycle, the analyst needs to regroup and collect his thoughts by first spelling out the content of the relevant information. Much of the information collected will be discarded as being of no further use. However, it is probably best not to throw away information that has been collected, until after the system study is completed. One's view of what is relevant and what is irrelevant often changes significantly during the course of the project.

In any case, the information in question needs to be organized into a meaningful form so that it can be analyzed for accuracy and for use in deriving other system study factors.

To see if the information really makes sense and thus determine its degree of usefulness, the analyst needs to know how it was derived, how reliable the methodology was that was used and, lastly, how valid the results are to the situation being studied (Huff).

Thus, the analyst first needs to determine how the basic information was derived. As we saw in chapter 9, the basic methods for gathering information are review of literature, interviewing, sampling, observation, forecasting, modeling, and testing. Each of these basic methods has its particular strengths and weaknesses. The analyst needs to check into the techniques used in a given case and assess the potential correctness of the reasoning that was followed.

How reliable is the methodology used as a means for generating consistently useful information? Is the analysis based on a biased sample which involves only a few cases or is in other ways unrepresentative of the true situation? Is a certain factor assumed to be the cause of an event, when in fact the evidence suggests merely a high correlation? Farmer's Insurance Company came out in the 1970's with an auto insurance policy which gave a discount to drivers who did not smoke. They had found that drivers who didn't smoke had fewer accidents than drivers who did smoke. How did they reach this conclusion? Is it based on the average or is it the result of high correlation or what? Assume it was the result of a study which showed a high correlation between the amount of smoking a person does and the number of car accidents he is involved in. Does high correlation prove that if people who smoke now will stop smoking, they will have less car accidents?

In arriving at the conclusions (information) presented to you, what basic assumptions have been made? Are these assumptions reasonable in the face of reality or was the problem merely rationalized to make for a mathematically tractable formulation?

Additionally, the analyst needs to consider the validity of the results with respect to the situation he is analyzing. This can be accomplished by comparing the information gathered to what was expected, based on the analyst's knowledge of the field. His expertise gives him a general body of knowledge which he can use as a comparison base. Further, he needs to cross-validate the results with other sources to see if similar results have been determined.

For example we look for incomplete truths, which lead us to a conclusion from the data given that really is not warranted. For example, on a TV commercial shown for Bayer aspirin, it was stated that a recent U.S. government study had shown that there was no aspirin more effective than Bayer's. Now most of us would thus feel very good about the effectiveness of Bayer's aspirin and therefore would be willing to spend a little more money to buy it than an unknown aspirin. That conclusion is wrong! What the U.S. government study actually showed was that all aspirins are alike (i.e., they could find no difference in effectiveness). Was Bayer lying? Isn't it true that if all aspirin are alike, then there is none more effective than Bayer's?

Frequently you will not have the time or expertise to evaluate fully some specific information. What can be very helpful is to do some quick "back of the envelope" calculations. You attempt to see if the figures shown are in the same ballpark (similar in magnitude) to those you get through very quick and gross calculations. For example, you could be presented the printout results from a very sophisticated computer-based mathematical model. Without going into the full ramifications of the model (which could take a month if you had the mathematical background), you can compare the results with possibly some historical industry ratios, or you could assume a simple algebraic formula like the ROI. You could take a sample of the total results to figure in detail, etc.

Especially if you don't have a way to evaluate the results directly, it can be very helpful to determine what kind of different results can be derived by slightly changing the assumptions or estimates used. A determination of the validity of the most sensitive assumptions or estimates could then be made. You could then concentrate on those areas in which the final results are most sensitive and which there is the least amount of confidence about the data. Depending on the time available and worth of the results, further efforts could be made to cross-check these points.

EVALUATING SOURCES OF INFORMATION

In evaluating the quality of information, the analyst needs to judge the competency and credibility of the source. Which factors to use and how to proceed depends on whether the source is a person or an organization. To see how this works, let's first consider how to judge a person's competency.

Judging Personal Competency

If the information has been derived from a person, the analyst needs to judge the competency of that person to speak on this particular subject matter. As shown in Table 11-3, this judgment should consider a person's general ability, specific expertise, and conviction (Prell).

General Ability: The overall ability of a person influences their competency as a source of information. This ability can be measured by considering a person's intelligence, education, clearness of thought, and awareness and judgment.

The more intelligent a person is, the more likely you should value their ideas. Some aspects of intelligence are measured by IQ tests, Scholastic Aptitude Test scores, Graduate Record exams, etc. But beyond this, we need to consider a person's general curiosity to learn, or desire to know more about how things work, etc. A most important aspect of intelligence is the ability to view situations from a variety of perspectives.

If intelligence was all that was required to know what one is talking about, we should all follow the "geniuses" of the world. But we know there are other facets of general ability which generally are more important than "IQ." What kind of formal and informal education does the person have? What kind of schooling has the person completed, such as college, company schools, apprentice training, etc.? We need to

Table 11-3 Judging the competency of a person.

General Ability	Specific Expertise	Conviction
Intelligence	Formal education	Attitude
Education	Experience	Degree of confidence
Clearness of thought	Track record	
Awareness and judgment		

establish the level attained, the quality of schooling, and how well the person actually did.

In general, everything else being equal, the more formal education a person has, the more competency he has on average. For example, a person should be more learned if he has a Ph.D. than a Master's, or is a college graduate versus a high school graduate. But it is not only a question of the level attained, but also where the education was attained. In general, a person would have learned more at one of the prestigious universities than he would at a lesser known local college. In turn, with some of the finer high schools it is possible that a person trained there would have learned more than a person educated at a four-year college. A school's reputation for excellence should be an important indicator.

How well the person did at these various schools should also be considered. Was he or she in the top ten percent, the middle, or the bottom of the class? The better the performance, in general, the greater the competency.

We have talked about formal education in terms of traditional schools like high school and college, but there are many other ways to gain a formal education in particular fields. The military has many excellent training schools in subjects like electronics, jet engines, cooking, medical technology, etc. Most major companies have training programs. For example, IBM and Xerox both have extensive programs in the computer field to keep their employees updated. In many of the crafts, there are apprentice training programs. Thus, when thinking of formal education, we needn't limit ourselves to evaluating only the more traditional institutions.

If formal education and intelligence were the only important indicators of determining a person's competency, then university professors would be the most knowledgeable. From your own experience in school, you know this isn't always true. There has got to be more to competency than just "book-learning."

A person's clearness of thought in both written and verbal form are usually very useful indicators of general ability. The factors of intelligence and education should highlight whether a person has something worthwhile to say. But here we want to measure how well they can get their ideas across. Can the person adjust his thoughts to various audiences? Is the person able to get across very complex thoughts to experts and novices alike?

The last major indication of general ability is awareness and judgment. Intelligence, education, and clearness of thought basically measure a person's potential. But equally, if not more importantly, we are interested in what a person has done and what the degree of success has been. While classifying people as winners and losers is too simplistic, it is a fact that some people can make things happen. Their record of success on important projects is very high. Whereas, some other people with apparently equal ability can't ever get the jobs done.

In summary, a person's general ability is a function of their intelligence, education, clearness of thought, and awareness of judgment These factors should not be viewed in an absolute sense; that is, a person with a Ph.D. always knows more than a person with a sixth grade education, the general always knows more than the private. Rather, these should be taken as probability statements. For example, the greater the education attainments of a person, the more likely she will be competent, everything else being equal. It is important to recognize that in our discussions we are taking one indicator at a time and showing how competency varies with different levels of this factor. It is assumed that all other factors are held constant. We will see at the conclusion of this section that we need to make an overall judgment of a person's competency and we have to consider all factors together.

Specific Expertise: The analysis of a person's general ability gives an indication of overall competence, but as an information source, we are primarily interested in the degree of expertise in the specific area in question.

Let's pose an example at this point to show the important difference between general ability and specific expertise. Many years ago, Dr. Benjamin Spock wrote a best-selling book on the topics of caring for babies. It rapidly became "the" reference book in this subject matter. Dr. Spock was a medical doctor who had spent many years practicing pediatrics and had a gift for explaining complex, medical concepts in a way that young parents could understand. By our indicators of Table 11-3, he had very high general ability and when he spoke on babies his specific expertise was excellent. Overall, he was a competent source of information.

Years later, Dr. Spock became an activist speaking out on the subject of nuclear reactors. How would you rate his competence in this area? Because he is competent on the subject of babies does not imply he knows anything more than anybody else on the nuclear situation—thus the reason for determining specific expertise.

What is a person's formal training in the specific subject matter? What educational level did he achieve, what was the quality of the schooling, and how did the person perform? In addition to education, what experience does the person have in his field? How many years has the person been in this type of work? What kind of experience is it, in terms of its similarity to the situation you are interested in? What is the quality of the experience in terms of the kinds of projects the person has been responsible for?

What is the person's record of success in this specific field? To determine the track record, you can check factors like numbers and types of success, the person's present job title and his promotion or advancement record. To a large extent, a person's salary in relation to others in his field is a good indicator of success. What recognition has the person been given by his peers, superiors, or subordinates? The quality of the awards, medals, etc., is important here.

Conviction: We have considered how general ability as well as specific expertise in the subject matter area serve as indicators of a person's competency. But we haven't covered how competent the person feels about the specific questions or project that we are interested in.

What is the person's conviction that what he is saying is the way it is? How confident is he as to the accuracy or correctness of his statements? In the later part of this chapter, we will discuss the details behind establishing an "Overall Confidence Indication." At this stage, we will relate the point to watching the TV news for a forecast of tomorrow's weather. The weatherman says it is going to rain tomorrow. We would like to know his confidence in saying that. In other words, is he 100% sure, or is it a 50-50 chance, or is it really a 10% change of happening?

In judging other people's degree of confidence, we need to assess their general overall attitude. This is a base which greatly impacts on how one sees specific things. For example, is the person usually optimistic or pessimistic about life in general? The optimist sees most things as "can do"; if there is a problem, it will work out. The pessimist approaches most situations by the "Murphy's law" method—if anything can go wrong, it will. On the other hand, there are people whom we will call realists. They view each situation on its merits, and withhold judgment until they have assessed the facts.

Summary: When we don't have the expertise to evaluate the information itself, we have to judge the competency of the source. If the source is a person, we should determine this individual's general ability, their specific expertise, and their conviction regarding the accuracy of the information they are giving.

These factors need to be viewed in a probability sense. Further, it is clear that one can't always find out all the relevant aspects of an individual. Other times, it isn't really worth it to spend the time necessary. Nonetheless, we now at least have a way of judging the competency of a source and can adjust our approach to the importance of the problem.

As an example of how to apply these thoughts, assume you were having a difficult time growing a grass lawn at your house. Because you have no expertise yourself (i.e., tried and failed), or you don't have the time or inclination to study how to grow a lawn, you decide to rely on evaluating various sources of information.

Over a barbecue dinner, you ask your neighbor what is the best way to get a good lawn. This guy is thrilled you asked and proceeds to tell you all he knows plus more. In trying to evaluate his competency, we aren't interested, per se, in whether he has a doctorate in physics or a high school education in auto mechanics. What we want to know is, What his level of skill is in the growing and caring of lawns? How many lawns has he taken care of? How many has he grown in your particular type of climate, soil, etc.?

Most importantly, what is his track record? How does his present lawn look like right now? If it looks brown and full of weeds, better seek council elsewhere even though he talks a good game. So we walk around the neighborhood and seek out the guy who has the best-looking lawn. Even though this person may have little formal education, he has the track record. We listen to how he attained success.

Judging Person's Credibility

We have given examples of factors to use in judging a person's competency as a source of information. Implicitly it was assumed that a person who knows what he is talking about would be willing to tell us all he knew. Unfortunately, for reasons we will go into later, many people like to play their cards "close to the vest" instead of revealing all. Additionally, there can be tremendous group or organizational pressure on a person to keep from saying what he believes or feels. In situations like these, the person doesn't really tell you what he really believes, and therefore his overall competence is undermined.

Thus a person's credibility becomes an important factor along with competency in evaluating the source of the information. In Table 11-4, the major factors affecting credibility are grouped under "Inherent Character" and "Situational Pressures."

Table 11-4 Judging the credibility of a person

	Inherent Character	*Situational Pressures*
	Candid	Social
	Consistent	Psychological
	Considerate	Setting
	Integrity	

Inherent Character: The underlying character of a person can be judged by looking at factors such as how candid the person usually is, how consistent she is in expressing a viewpoint, how considerate she is of others, and basically what the integrity of the person is (Prell).

The more candid a person is, the greater the credibility. This can be measured by sensing how frank, open and sincere, or straightforward a person has been in an interview situation. For example, you could ask the person a question about which you know he will be uncomfortable, and see how he reacts.

Can the person talk about the mistakes he has made along with the successes? When talking with a person, do you feel he is saying what he feels or is there some kind of hidden agenda or ulterior motive?

How considerate is the individual of others? Can she empathize with others by putting herself in their shoes? How tactful and courteous is she in explaining delicate matters?

Observe how the person reacts to you. Does she talk down to you? Is she arrogant as opposed to confident? Does she make you feel foolish when you ask questions, or naturally comfortable?

Does the person take a position in regard to an issue, or does she tend to "blow with the wind"? Telling us what we want to hear makes us all feel better, but it significantly lowers the person's credibility. If every time you talk to a person, she has a different story or rationale, this inevitably reduces her credibility.

Oftentimes what a person tells you is what she feels, but she does not tell you the whole situation. By sensing what topics the person doesn't want to get into and by sensing ideas which are missing or not said, the analyst can read "between the lines" to judge the person's credibility.

The final factor in estimating the character of a person is her integrity. Is the individual a person of principle? Is she responsible and does what she has agreed to do? In tough situations, does she show impartiality and make a judgment on the merits of the case as opposed to being out to get someone? The less honesty and fairness a person shows, the lower the degree of credibility.

As with the factors used in judging competence, the qualities of a person's being candid, consistent, considerate and showing integrity should be used in a probability sense. Additionally, each factor is continuous. A person is not entirely honest or dishonest, but rather has a certain degree of honesty. All these aspects combine to indicate the inherent character of a person.

Situational Pressures: As we have seen, in judging a person's credibility we need to first investigate the underlying or inherent character of that individual. Unfortunately, that assessment isn't enough, because situational pressures can cause a person to be less credible than he "normally" is.

The first thing to note is whether you are asking the person to speak as an individual or on behalf of an organizational group. The more one is speaking as a member of an organization, and especially the larger the organization with the longer "chains of commands" to approve comments, the more the pressure for the person to speak as the company would want him to speak. This in turn lowers his credibility since he most likely won't be speaking as he feels.

Whether a person is speaking as an individual or as part of an organization, you need to assess the relevant reward or penalty structure. That is, does the person have something to gain or lose by what he says?

One does not expect the Chevy salesman to tell you that the VW Rabbit is the best car for you. But how about a college professor telling you to switch majors from his field to another one? Are there any rewards or penalties attached to what either the car salesman or college professor has to say? Yes, there are. In the case of the car salesman, he is on a commission basis so he loses when you go buy a Rabbit as opposed to a Chevy. The college professor is less directly affected, so we should expect a more truthful answer.

Is the setting friendly or hostile? By that we mean, in the case of the interview, does the interviewee have a "chip on his shoulder"? The friendlier the setting, the more likely a person will say what he feels. The brusque, hard-headed interviewers who play the game to win regardless of the rules of courtesy, respect, and tactfulness necessarily cause ill will. This generally means a person will be less likely to open up in this type of situation.

Among ways of easing the situational pressures, we could try to gain the confidence of the person being assessed. This can be done by showing empathy for the underlying situation, e.g., backing off when you see by a person's tone of voice, facial expression, body language and other nonverbal symptoms that he is becoming very sensitive to what is being said.

Listen not only to what is said, but how it is said. Frequently one can say to an organizational person that the discussion will be kept "off the record." Taping of conversation oftentimes makes people nervous because of the permanency of the recording.

In sending out questionnaires, one should allow the respondents to be anonymous. As a minimum, they should be guaranteed their specific answers will be considered confidential.

In summary, the greater the situational pressures, the less likely a person will say what he really believes. Where possible, the analyst needs to reduce the social and psychological pressures and develop a friendly setting, to insure a high degree of credibility.

The importance of credibility in the equation of judging overall competency of a source should not be minimized. As former President Nixon found out, even though he had a very high level of competence regarding presidential matters, when he lost his credibility he was eliminated from office.

Judging an Organization as a Source

In the first two parts of this section on evaluating the source of information, we have dealt exclusively with people. And while it is true that people are ultimately the source of all information, sometimes we can't determine who the person really is. Additionally, the person may not be speaking for himself, but rather for the organization. In any case, there will be many instances when we will need to judge the usefulness and accuracy of information produced by an organization.

Basically, we will apply the same scheme as used in judging the competence and credibility of a person, only we will need to modify some of the subcategories to organizational factors. In determining the general ability of an organization, we can look at factors such as gross sales, profit, market shares, and years in being.

Evaluating specific expertise would involve checking out a company's reputation for excellence, awards they have been given, number of projects they have completed in the area of the subject matter under study, percentage of sales in their area, etc. Additionally, we could check the competency of the people they have working for them in the field under study.

The area of conviction would entail determining the general attitude of the company. Is it generally optimistic, pessimistic, etc., toward the future? What is the degree of confidence expressed on the subject of the analysis?

The credibility of an organization can be seen by looking at its inherent character. How does it treat its people? What is its attitude toward customers? How fair and equitable have they been in past dealings? What is the company's public image? What kind of advertising does it use? Situational pressures, on the other hand, concern whether the company has something to prove, that is, something to gain or to lose by what they are saying. Further, is the atmosphere in which the information is being presented friendly or hostile?

As an example of organizations as a source of information, consider the recent "oil crisis" in 1979. It was not at all clear whether in fact there really was a crisis. The American people had to make a judgment about the situation based on information given by the press, oil companies, Congress, Ralph Nader, etc. While the oil companies most likely had the competency to speak on the subject, their credibility was questioned by most people, if for no other reason than they have something to gain or lose from the situation.

When Congress addressed the issue—or even the Department of Energy—it became clear they didn't have much competency on the subject matter because they stated they had no information per se, but had to rely on the oil companies to supply them with the required data. Further complicating the issue was the overwhelming complexity of the situation.

Additionally, the cartel of oil petroleum exporting countries was trying to present a united front on this question, which in turn put tremendous situational pressures on any individual member of the OPEC board who wanted to speak his own mind.

SUMMARY OF HOW TO EVALUATE INFORMATION

In this chapter, we have considered why it is necessary for the analyst to have to evaluate information. We noted that in most circumstances, the analyst would have to *both* evaluate the information itself and the source of the information to determine its probable accuracy.

How to do this is summarized in Table 11-5. What the actual information is needs to be spelled out and summarized in a meaningful form. To determine if the information makes sense, the analyst needs to determine how it was derived, how reliable that methodology is, and whether the results are valid in terms of the system being studied.

In evaluating the source of the information, whether that might be a person or an organizational group, the analyst needs to check the overall competence and credibility of the source.

The competence of the source is related to general ability, specific expertise, and conviction. Credibility involves establishing the inherent character of the source and the relevant situational pressures.

The continued evaluation of the information and its source results

Table 11-5   How to evaluate information.

What is the information?	*Does the information make sense?*
Spell out content	How was it derived?
Summarize into a meaningful form	How reliable is the methodology used?
	How valid are the results with respect to your situation?

What is the competence of the source?	*What is the credibility of the source?*
General ability	Inherent character
Specific expertise	Situational pressures
Conviction	

in an overall confidence indication as to the accuracy of the information. The explanation of the confidence levels follows in the next section.

OVERALL CONFIDENCE INDICATION

Since by necessity the analyst will not be able to verify the correctness of all the information he must gather in doing a system study, and further since much of the information needed is based on assumptions, guesses, and conjecture, the analyst needs to give an indication of the probable accuracy of the information.

This indication, while basically subjective, should be dependent on how the information was derived, how reliable the methodology is, and how valid are the results in regard to the system being studied. Further, the competence and credibility of the source of the information needs to be established. As we have seen, this can be determined by considering the person's general ability, specific expertise, conviction, inherent character, and the situational pressures.

Codes

To aid in the use of this concept of confidence levels, we will define four categories, as shown in Table 11-6.

A "Very Confident" rating means the analyst feels that greater than 80% of the time this information would check out as stated. For example, if the weatherman says it will rain tomorrow, what kind of confidence do you have in his opinion? Assume you followed his forecasts for 100 days. How many times was his forecast of rain/sunshine/morning drizzle, etc., correct? If it was 62 times out of the 100 forecasts, we should be "Confident" in what he has to say.

Table 11-6 Overall confidence indication.

Confidence Indication	Code	Percent	Factor
Very confident	VC	> 80%	.9
Confident	C	50-80%	.6
Little confidence	LC	20-50%	.3
No confidence	NC	< 20%	.1

As with all things, determining confidence is not this simple. When is precipitation classified as rain? What is the specific difference between light showers, drizzle, downpour, etc.? Actually we needn't dissect the confidence concept too finely. Our purpose is just to get an overall feel for the accuracy of the information and specify your conclusions into four broad categories. We certainly don't want to try and say we are 87.32% confident. . . .Moreover, as arbitrary as these categories of confidence are, they surely are better than not broaching the subject and thus implicitly assuming that information is always 100% accurate.

The factors of .9, .6, .3, and .1 will be used in the evaluation matrix explained in chapter 12. For certain purposes, we will want to discount information depending on its potential accuracy. These factors are approximately the midpoints of the various ranges, i.e., VC is 80% to 100%, so the factor is 90% or .9, etc.

To better see how the concept of confidence can be applied in a system study, we now consider two examples. The first concerns the overall potential accuracy of estimating the gas mileage a car would get using various sources of information (Table 11-7). The second example derives confidence indications regarding the means used for estimating inflation over the next four years (Table 11-8).

Determining Gas Mileage Example

In Table 11-7, several different methods for gathering information are represented, and they are rated as to their potential accuracy. The first example shows an estimation process which should yield at least an eighty percent chance of being "right on." Consumer Union is a highly competent source as a testing agency. It has developed much expertise in this regard over the years and has a good track record for accuracy. Its credibility is high since it accepts no advertising and thus reduces situational pressures. Further, this organization tells both the good and bad points of a product. The testing procedures seem to be very reliable since they cover many traffic situations and have been applied to many cars of the same model. Further, the validity is good since the model and its configuration (i.e., Toyota, Corolla, 1600 cc engine, automatic transmission, etc.) are the same in this case as we are interested in buying.

But even as good as this is, it is rare that one source by itself should yield a VC rating. By cross-checking with the estimates of other major agencies such as the government, an automobile association, and a consumer car magazine, your confidence level can be increased if the gas mileage estimates are all very close. Also note that these three ad-

Table 11-7 Determining a car's gas mileage.

Code	Percent	Examples of Information Sources
VC	$> 80\%$	(1) Major independent testing agency like Consumer Union has run extensive tests among many of the same car models over various traffic situations (freeway, cross-country, stop and go). These findings are in close agreement with the estimates of the Environmental Protection Agency, Southern California Automobile Association, and *Hot Rod Magazine.*
C	50-80%	(2) Consumer Union, EPA, and the Southern California Automobile Association gas mileage estimates for the various car models agree within 3 to 5 mpg.
		(3) A semi-independent car magazine like *Motor Trend* or testing agency (commissioned by Ford Motor Co.) concludes after running various tests over several traffic conditions using a couple models of each car that. . . . These findings are in close agreement with your personal driving comparison.
LC	20-50%	(4) The estimates of the major independent testing agencies show wide disagreement (6 to 10 mpg) of what various car models will get for gas mileage.
		(5) You got your estimates from a cross-country gas economy run sponsored by a major oil company (i.e., Mobil Oil's L.A. to New York City gas economy run for new cars).
		(6) Upon inspection of the sales brochure at the car dealers, you note some mileage estimates by the manufacturer for a particular model (i.e., the Cadillac gets 15.6 mpg).
NC	$< 20\%$	(7) You have concluded that the Datsun gets close to 40 mpg because you heard on a TV commercial that one got 39.4 mpg on the trip from Washington D.C. to Kennebec, Maine.
		(8) A new car salesman, in trying to sell you a Gremlin, states the car gets well over 25 mpg.
		(9) Your buddy (who tells you he always wins at Las Vegas) states that his Buick Roadmaster gets 21.6 mpg.
		(10) You drive a demonstration car around town for several hours and estimate the mpg from gas gauge readings.

ditional sources are from different industries and thus have different motivations and methodologies. This is generally better than getting four estimates all from car magazines.

In source example 2, the same sources are used, but there is a 3-to-5 mpg difference in gas mileage estimates. Due to this variance, one should be confident of an estimate as opposed to very confident. Source example 3 has only one major source. The source *Motor Trend* as a testing agency has a high competence level. However, the potential credibility must be lowered since *Motor Trend* accepts advertising. This implies that all other things being equal, there can be situational pressure on *Motor Trend* to go easy on a car it is reviewing, if that manufacturer advertises heavily in their magazine. Note, however, we are not saying this will happen, only that the probability exists. The same is true with the testing agency commissioned by Ford Motor Company. For any follow-on work, the testing agency most likely needs to determine good results for Ford's car. However, since these "conflict of interest" aspects are only potential and because Ford and *Motor Trend* have very good track records and their findings are in close agreement with your personal driving comparison, a confidence rating of C is given.

Source example 4 is the same as 1 or 2, but there is even wider disagreement among the results, so the information is discounted to a 20% to 50% confidence factor or LC rating. In source example 5, there is only one testing procedure. Further, it is sponsored by an oil company which has something to gain from the results. For example, if all the new cars in the test got far better gas mileage than people usually get, then people would probably conclude they are using the wrong gasoline and should switch in this example to the Mobil brand. While the test seems more reliable than the usual dynometer reading, since the cars are driven under typical conditions for over 2500 miles, the results don't make any sense. That is, you can't cross-validate them with other mileage estimates that give mpg as 10 to 15. The Chrysler New Yorker got 29 mpg; the Ford Van got 27 mpg, etc. While you might not be able to determine what the problem is, you should greatly discount the results because of their lack of validity.

Inspecting the sales brochure is not a very reliable way of determining gas mileage. On the other hand, it might be the only source of information available. So we always use what we can, while taking into account its potential accuracy. Since the Federal Trade Commission has cracked down on misleading advertising, the 15.6 mpg estimate in the Cadillac brochure is probably accurate. That is, they did under certain circumstances get this mileage. The only problem is that the conditions aren't close to your driving situation and thus the validity is questionable. Further, the brochure is a sales tool and thus needs to be discounted. In spite of these drawbacks, there is probably a 20 to 50% chance of accuracy with this information.

Source 7 is very low in potential accuracy since it is just one person driving in one situation. Further, it is a TV commercial for which the company would not be spending money unless the results were favorable and thus the chance for selecting only the "good" tests. Lastly, there hasn't been any testing agency certifying the results. New car salesmen are not known for their high credibility. Source 8 shows a salesman whose standard of living is dependent on selling cars, so there is a high likelihood that he will slant the information in a way favorable to making the deal. Further, nothing is put in writing; it is just a casual comment.

Source 9 is a single-source situation from a buddy of yours who has overall low credibility. Further, the results don't make sense. Lastly, in source 10, you drive a car around town for several hours. This test is not very typical of your driving situation, it only involves one car, and the gas gauge is not a very reliable instrument.

For sources 7, 8, 9, and 10 you would have less than a 20% chance of accuracy with these methods. You might ask at this point why even consider these types of sources at all. Most often this is the best you can do because of limitations of time, money, etc., or the overwhelming complexity of a problem, etc. How accurate do you think your information sources and methods are for determining a career or selecting a mate? Thus, one has to do the best one can by using whatever information sources are available, while always taking the nature of the sources into account.

Determining Inflation Rate Example

Although the first example of gas mileage may seem relatively simple and straightforward, this next example of estimating what the inflation rate will be is much more complex. The underlying factors are many, they are not well understood, and furthermore we are dealing with a future time period subject to much environmental change. In spite of these complications, the evaluation of the sources for potential accuracy follows the same procedures as in the gas mileage example. Thus, we won't go into as much detail on the explanation and interpretation.

In source example 1 (Table 11-8), a person should be very confident that more than 80% of the time when the President's Economic Advisors, Chase Manhattan Bank, Wharton's School of Finance and General Electric all come up with estimates which are are quite similar, this is the way it will turn out. These are major sources enjoying reputations for high competency and, in general, good credibility. Fur-

Table 11-8 Determining the future inflation rate.

Code	Percent		Information Sources
VC	> 80%	(1)	Reviewed studies done by prestigious major organizations, councils, like the President's Economic Advisors, Chase Manhattan Bank, Wharton's School of Finance and General Electric and almost all forecasts are very similar.
		(2)	Reviewed studies done by major organizations (Chase Manhattan, Wharton, G.E., etc.) and there is a majority indication around a particular growth rate.
C	50-80%	(3)	Reviewed lesser prestigious organizations, economic forecast estimates like *Fortune* magazine, UCLA Graduate School, Security Pacific Bank, or Standard and Poor.
		(4)	Paul Samuelson, Professor of Economics at MIT and Nobel Prize winner, feels the inflation rate will be . . . .
		(5)	Reviewed studies done by major organizations and there is wide variation in their estimates.
LC	20-50%	(6)	The economic advisor to your local newspaper, states in her weekly financial column that the rate of inflation will be . . . .
		(7)	You were given a free estimate by a state college professor in the school of business.
		(8)	A buddy who is an economic major states that . . . .
NC	< 20%	(9)	You overheard a bank clerk at the Bank of America say . . . .
		(10)	Your father-in-law, who is a plumber, tells you . . . .

ther, they are highly different sources. However, why aren't we 100% sure? The President's Economic Advisors' track record over the 1970's in estimating inflation has been poor. They have been "wrong" many times.

Source 3 shows that there is a different competency level among organizations. Security Pacific Bank is not in the "class" of Chase Manhattan. In source 4 and source 6, if Paul Samuelson speaks, we would listen. While he is not always right, he does have the track record and a greater competency level when it comes to predictions on inflation than say Sylvia Porter and her syndicated financial column in a local newspaper. This in no way implies that we can't learn from

Sylvia Porter. It just says that in a probability sense we would have lower confidence in what she has to say about future economic trends than Paul Samuelson.

In both of these examples of gas mileage and inflation, it is important to consider that the confidence indications were given as probability estimates, based on available information. It may well be that your father-in-law, source 10, turns out to be more correct on inflation *over the years* than does the President's Economic Advisors. If that is so, the respective track records should change your probabilities so that you are more confident of your father-in-law's estimates and less confident about the President's Economic Advisors.

Class Exercise 11-1

WHAT IS YOUR EXPERTISE?

In the everyday problems we all have to face in life, I would say that in 80 to 90 percent of the situations we have little or no expertise. Test the validity of this statement in the following situations.

1. Taking your car into an auto repair shop for a transmission problem. How do you know if the problem was fixed correctly and at a fair price?

2. A medical doctor says you have a serious stomach problem which requires major surgery. How do you determine if he is right?

3. Determining what career is best for you. That is, how do you know you will like being a dentist (i.e., accountant, lumberjack, etc.) prior to actually being a dentist?

4. How could you get expertise in these previously mentioned situations? How could you evaluate the sources?

5. Is there any relationship between our lack of expertise and some government agency requiring doctors to be certified by a medical board, airline pilots to be licensed, establishing a consumer protection board, etc.? Is this really necessary? How about the desirability?

6. How do you know a book will be worth reading, prior to actually reading the book? That is, if you can't judge a book by its cover what can you do to increase your chances of reading good books? Go to the local bookstore and assess the various books according to criteria such as who is the publisher, reviewer's comments, the look of the cover, etc. Which are valuable predictors and which aren't?

Class Exercise 11-2

REQUEST FOR PROPOSAL

One of the major ways by which companies procure computers is to send out a request for proposal (RFP). The RFP lists the company's requirements in terms of jobs it needs to accomplish. Vendors submit proposals for computer equipment and its accompanying applications software to meet the company's data processing needs. Assume the controller for a small business has prepared an RFP to cover their accounting needs over the next five years.

1. Discuss the potential accuracy of the company's requirements as spelled out in a RFP. What is the confidence level for today's requirements versus those projected for a future time period?

2. How would you establish the accuracy of the proposals submitted by the vendors? How would you establish their competency and credibility?

3. DATAPRO (i.e., the Consumer Union of the computer field) surveys users of various computer equipment. This nationwide survey tabulates responses from tens of thousands of users. Some performance ratings for a specific vendor are based on a sample size of 3, others 25, and others 300. What is an adequate sample size for you to be confident of the results? What do you do if the survey gives a vendor excellent marks by 200 users on the XYZ Model B computer and you are thinking of buying the newest XYZ Model C computer.

Class Exercise 11-3

SHOULD THE BUILDER ADD AIR CONDITIONING?

Assume you are a homebuilder who is thinking about building a housing tract in the Mission Viejo (California) area. One of the factors in your decision on what kinds of houses to build is whether they should have central air conditioning. Because you want the homes to be as affordable as possible, but at the same time comfortable as necessary, you need to establish how hot it gets in this area. Because you had never lived in southern California before and it was December when you were looking, you decided to check with various information sources on how hot it gets during the summer and thus determine whether air conditioning is essential.

Information Sources

1. You called the local air conditioning sales outlet and talked to the owner about whether air conditioning is needed or not. He states it certainly is a must to live comfortably, because during many days in summer the temperature gets over 100°.

2. You talk to a real estate agent who is representing a homeowner whose house is without air conditioning. He says that some people have air conditioning, but most nights you will need covers to sleep in. So air conditioning is not really necessary.

3. You go to the local chamber of commerce and pick up a brochure about the city. The brochure states that the mean year around temperature is a cool, comfortable 72°.

4. You talk to a potential neighbor who has lived in the area for 4 years. This grandmotherly woman says they have never even needed a fan. On the particular day you talked to her it was 85° and while you were sweating profusely, she talked to you the full 30 minutes with a winter overcoat on.

Table 11-9 Climatic Study El Toro area.

CLIMATIC AREA NUMBER 03
SOUTH COAST DRAINAGE

EL TORO MCAS, CALIFORNIA

LATITUDE 33.40 N
LONGITUDE 1177.44 N

ELEVATION 393 FEET MSL

The following is a climatic study of the El Toro area. The study uses 19 years of actual weather information.

PARAMETER DESCRIPTION	JAN	FEB	MAR	APR	MAY	JUN	JUL	AUG	SEP	OCT	NOV	DEC	ANN
ABS MAX TEMP (F)	92	88	87	91	97	103	100	101	106	107	97	93	107
MEAN MAX TEMP (F)	63	65	67	69	72	76	81	81	81	76	72	66	72
MEAN MIN TEMP (F)	44	45	46	50	53	56	61	60	59	55	50	46	52
ABS MIN TEMP (F)	25	32	33	33	39	45	48	47	44	38	35	30	25
MEAN NO DAYS TEMP-OR GTR 90 (F)	0.1	0.0	0.0	0.1	0.4	0.6	1.7	2.1	3.8	2.0	0.7	0.1	11.6
MEAN NO DAYS TEMP-OR LESS 32 (F)	1.4	0.1	0.0	0.0	0.0	0.0	0.0	0.0	0.0	0.0	0.0	0.4	1.9
MEAN REL HUM (PCT)	69	67	70	73	71	73	73	75	72	71	62	63	70
MEAN PRECIP (IN)	3.62	2.09	2.00	1.63	0.29	0.02	0.01	0.04	0.10	0.26	1.01	1.69	12.8
MEAN NO DAYS PRCP-OR GTR 0.1 (IN)	5.9	3.5	3.6	2.7	1.1	0.0	0.0	0.2	0.2	0.6	2.1	3.2	23.1

DEFINITIONS:
ABS MAX = Absolute maximum in 19 years.
MEAN MAX = Average maximum in 19 years.
MEAN MIN = Average minimum in 19 years.
ABS MIN = Absolute minimum in 19 years.
MEAN NO DAYS TEMP = OR GTR 90 (F) = Average number of days the temperature is equal to or greater than 90 degrees Fahrenheit.
MEAN NO DAYS TEMP = OR LESS 32 (F) = Average number of days temperature is equal to or less than 32 degrees Fahrenheit.
MEAN REL HUM (PCT) = Average relative humidity in percentages.
MEAN PRECIP (IN) = Average rain fall in inches.
MEAN NO DAYS PRCP = OR GTR 0.1 (IN) = Average number of days rain fall is equal to or greater than one tenth of an inch.

5. You made a quick survey of this area and noted that of the 10 middle-class houses you looked at, 7 had air conditioners.

6. You went to the local airport which is just 10 miles the other side of the Mission Viejo mountain range, and got from the Weather Service the official data in the accompanying table (Table 11-9).

Questions

(1) What is the competence and credibility of each information source?

(2) Rank order each source of information as to your overall confidence level.

(3) What other sources of information would you want to use?

(4) Should the builder add air conditioning?

Chapter Twelve

The Evaluation Process

In the first eleven chapters of this text, we have introduced what the systematic systems approach is, how it relates to cybernetic systems, and further how to formulate problems, develop system solutions, establish which alternatives to evaluate, then how to evaluate alternatives, how to gather information, what information to gather, and how to evaluate information. While we still have many more ideas to cover before you see all the aspects of the systematic systems approach, it is time to collect our thoughts and see the whole picture.

Figure 12-1 shows an overview of the systems approach to problem-solving. There are ten major processes involved in the total approach. Each of these processes requires certain inputs to work on and in turn each process produces certain outputs or results. We have covered to this point in the text, processes 1, 2, and 3. We have discussed in detail the inputs/outputs of symptoms, potential system designs, constraints, feasible alternatives, and criteria.

In this chapter, we are going to cover process 4—system simulation, 5—preference chart, 6—systems utility function, and 7—evaluation

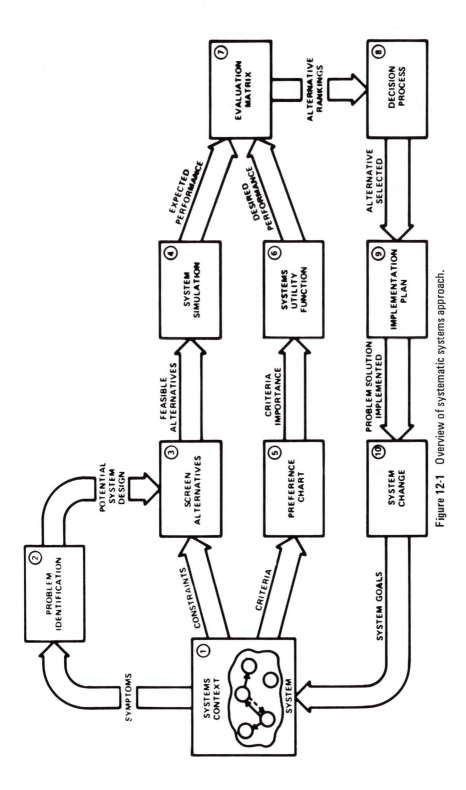

Figure 12-1 Overview of systematic systems approach.

205

matrix. Further, the appropriate inputs and outputs of expected and desired performance, criteria importance, and alternative rankings will be discussed. In the remainder of the book we will cover the decision process (chapter 13), and implementation plan (chapter 14).

PREFERENCE CHART

At this point, assume that from the study of the systems context and much discussion within the problem-solving group, has come a set of criteria. These criteria have been judged to be useful dimensions on which to rate the relative worth of potential system designs. Surely, however, while all these criteria are considered crucial to the decision, some are more important than others. Establishing the relative importance among twelve or eight or even five criteria is generally a very difficult task. One method for aiding in this regard is the *preference chart* (see Fig. 12-2).

In the particular example, a student is trying to determine the best car for his family system upon graduation from college and entering the business world as a marketing representative. After much thought about his personal system and what it is trying to accomplish and the fact that there will just be one car for himself and his wife, they have determined some eleven criteria as being important to this decision—namely, cost, roominess, safety, etc.

Each criterion is shown along the left side of the chart running vertically. Additionally, the same criteria are shown across the top of the chart running horizontally. Each criterion is compared with all other criteria, one at a time. In each comparison, the analyst and decision-makers are trying to decide if the criterion on the left is more or less important than the comparison criteria on the right. More specifically is it much more, more, same as, less, or much less in importance. This is indicated in the chart as follows:

Symbol	Meaning	Points
$\gg$ or ++	much more	4
$>$ or +	more	3
= or 0	equal	2
$<$ or −	less	1
$\ll$ or − −	much less	0

Figure 12-2 Preference chart.

Criteria	Cost	Roominess	Safety	Style and looks	Comfort	Gas mileage	Air conditioning	Resale value and likelihood to sell	Frequency of repairs	Sound system	Warranty and reliability of service	Numerical value	Importance Grouping	Weighting
Cost		∧	=	=	=	∧∧	∧	∧	∧∧	∧	∧	28	Very important	3
Roominess	∨		=	=	=	∧	=	∧	∧	∧	∧	24	Important	2.4
Safety	=	=		∧	=	∧	=	∧	∧	∧	∧	27	Very important	3
Style and looks	=	=	∨		∧	∧	∨	∧	∧	∧	∧	24	Important	2.4
Comfort	∧	∨	=	∨		∧	=	∧∧	=	∧	=	23	Important	2.4
Gas mileage	∨∨	∨	∨	=	∨		∨	∨	∨	∨	∨	10	Minor important	1
Air conditioning	∨	=	∧	∧	=	∧		=	=	∧	∧	23	Important	2.4
Resale value and likelihood to sell	∨	∨	∨∨	∨∨	∨	∧	=		∨	∨	∨	9	Minor important	1
Frequency of repairs	∨	∨	∨	∨	∨	∧	=	∧∧		∧	=	19	Moderately important	1.8
Sound system	∨∨	∨	∨	=	∨	∧	=	∧	∨		∨	15	Average important	1.5
Warranty and reliability of service	∨	∨	∨	=	∨	∧	∨	=	∧	∧		18	Moderately important	1.8

*This particular format of the preference chart was developed by Nicholas Strauss.

After all the comparisons are made, a point total is determined for each criterion on the left of the chart. This total is entered in the numerical value column. The analyst then groups criteria having approximately the same numerical value. These groupings are then designated by terms like very important, important, etc.

The weighting value can be determined in many ways. One way is just to take the numerical values themselves as the relative weights. A second way, and the way shown in the example, is to give the lowest importance grouping a value of 1. To determine the other group weightings, divide their numerical value by the numerical value of the lowest group. This will then result in a relative weighting. A third approach is to establish that the overall weighting must add up to 100 (100 percent). To get the appropriate weighting among groups, one needs to find a divisor which will yield this total, while still keeping the relative proportion.

For the particular preference chart shown in Fig. 12-2, the criteria have been grouped and weighted. For this family, cost and safety are the most important factors by which to judge potential car alternatives. Next in importance are the criteria of roominess, style and looks, comfort, and type of air conditioning. Least important are resale value and gas mileage. In fact, safety is three times as important as gas mileage.

Note that to you gas mileage may be much more important than style and looks. But that is not the question. The question is, What is most important in buying a car for *this particular family?* If a case can be made in studying this family's systems context that the criteria selected make sense, then that is all that is necessary. Different systems have different needs and wants. Through the preference chart, we have a way of seeing what the CDM (through the analyst) considers important for the system.

As a second example of a preference chart, consider Fig. 12-3 which shows the relative importance of the criteria used to select the best typesetting process for a suburban newspaper. Recall the earlier discussion of this case in chapter 7.

To make the preference chart more useful, consider the following points. The criteria should be put on the chart in a random order. This is to try and eliminate any bias initially. Secondly, the two criteria at a time comparison should raise the following type of question: Is capital investment much more important, more important, about equal in importance, less important, or much less important than product quality to this newspaper? That is, of the two criteria which would the CDM give up some performance to get better performance on the other criteria?

We know the CDM really prefers to have much better product quality than could normally be expected, because this is a more-the-better type of criterion. Conversely, the CDM would prefer to pay

Figure 12-3 Typesetting preference chart.

Comparison Criteria / Criteria	Capital investment	Product quality	Personnel	Status of personnel	Page cost	Maintenance	Input	Versatility	Time	Numerical value	Grouping	Weighting
Capital investment		>>	>>	=	>>	>	>	>	>	30	Very important	5
Product quality	<<		=	=	>	>>	>	=	>	6	Minor importance	1
Personnel	<<	=		=	<<	>	<	=	<<	9	Moderate importance	1.5
Status of personnel	<<	=	=		=	>>	<	<	<<	7	Minor importance	1
Page cost	<	>	>>	=		>>	>>	<	<	22	Important	3
Maintenance	<<	>>	<	<<	<<		>>	>	<	20	Important	3
Input	<<	>	>	>	<<	<<		>	<<	12	Moderate importance	1.5
Versatility	<<	=	>	>	>	<	<		<<	10	Moderate importance	1.5
Time	<	>>	>>	>>	>>	>>	>>	>>		28	Very important	5

209

much less than one can normally be expected to pay for typesetting equipment, because this is a less-the-better type criterion. We have already established this in the discussion on performance measures. Each criterion has been discussed individually with all other things (i.e., criteria) held constant. Now we want to vary the other things (criteria) one at a time and see the resultant tradeoffs.

To be more specific, we have to ask what performance can we normally expect on this criterion. From the systems utility function (see Fig. 12-4), we have determined average performance. Let's look at the comparison again between capital investment and product quality. We see from Fig. 12-4 that in this typesetting situation, capital investment will normally be around $550,000 and the average product quality for these type machines will be mostly solid type faces, good appearance, eye-catching and practical color break.

Now the question is, Would we rather spend less (capital investment) and get worse product quality or would we rather increase capital investment to get better product quality? What is more important to this suburban newspaper? In the preference chart example of Fig. 12-3, it was decided that the CDM would accept much lesser product quality if he could reduce the capital investment.

The next point about doing the preference chart is that only the comparisons on the upper part of the diagonal need be done. The lower half below the diagonal is just the reverse. That is, the comparison between capital investment and product quality resulted in $\gg$. When you go to compare product quality with capital investment, it should be $\ll$. You can save much time in doing the chart in this way. However, by making all the comparisons, you would have a good cross-check on the consistency of your thinking.

When you have finished the chart, add the numbers in a horizontal direction for each criterion. Arrange the criteria into natural groups by similar total point scores. To make for ease of interpretation later in generating the evaluation matrix, get a relative weighting by dividing each group score by the score of the lowest group. Don't worry about being exact—simply round off. Call 2.85, 3. The purpose is not exactness, but rather a general feel for criteria.

Lastly, interpret the resulting importance groupings and see if they make sense. If they don't, go inside and look at the one-by-one comparisons and check for reasonability. Don't be afraid to change things. The chart is an aid to your understanding. Adjust but don't manipulate. That is, don't fake yourself out.

SYSTEMS UTILITY FUNCTION

Coming from the preference chart is a set of criteria which the CDM and others feel will give the best means for judging for effective system solutions. Further, these criteria have been grouped and weighted according to their relative importance.

What is now necessary is to specify the relative desirability of different levels of performance on each of the developed criteria. That is, for each of the relevant criteria, one will need to develop some explicit measures of performance. These yardsticks will show what to consider as exceptional, above average, average, below average, and barely acceptable performance on *each criterion.*

As was discussed in detail in chapter 8, to get the various measures of performance on a particular dimension or criteria, the analyst could consider a normal distribution. The mean or median values of performance would be an indication of what could be generally expected using the various relevant alternatives. That performance which exceeds 90 percent or better of the alternatives would be considered outstanding performance. That performance which is among the lowest 10 percent would be considered marginal or barely acceptable and so on.

Each yardstick of performance can also show a numeric scale which will be used later as a means for rating how well the alternatives to be evaluated do on a particular criterion. Arbitrarily, it is suggested a 10-point scale be used with 10 points being given for exceptional performance, 5 points for average performance, and 0 points for the minimum acceptable performance. A crucial point is that performance measures across all criteria should all be based on the same point value scale (in this case a 10-point scale).

As a more specific example of what has been discussed, consider the following tactical level decision. A small suburban newspaper firm wants to determine the best method for setting type for the daily printing of its newspaper.*

After considerable study of the organization and having many conferences with its management and key workers about the firm's present and future situation, the analyst felt the following system objectives were important in determining the overall effectiveness of the company: desirability of product to users, life cycle cost, reliability of process, growth capability, and employee morale.

Since the scope of the system study was limited by management to the tactical level decision as to the best way to structure the typesetting process, the overall company objectives needed to be tailored to this specific subsystem. This refining process results in the development of criteria. Criteria are the precise specification of the more generalized system objectives. For this company, the criteria that made most sense were: (1) versatility to customer, (2) product quality, (3) capital investment, (4) page cost, (5) personnel competence, (6) maintenance of system, (7) time element, (8) input volume, and (9) status of personnel. The criteria of versatility to

*This example is an adaptation of a system study performed by Barry Tharrington.

Weight	Rating / Criteria	0	1	2	3	4	5	6	7	8	9
			Barely Acceptable		Below Average		Average		Above Average		Exceptional
5	Capital investment		$800,000		$700,000		$550,000		$400,000		$300,000
	Time element		(a) 7 lines per minute (b) 10 hours		(a) 10 lines per minute (b) 9¼ hours		(a) 15 lines per minute (b) 8½ hours		(a) 20 lines per minute (b) 7¾ hours		(a) 30 lines per minute (b) 7 hours
	Page cost		$3.00 per page		$2.60 per page		$2.43 per page		$2.30 per page		$2.20 per page
3	Maintenance of system		(a) Company service to be called in as needed (b) Service when necessary		(a) School trained hot-metal with company men (b) Intermittent systems check		(a) Company machinist and company service available (b) Regular systems check		(a) Company machinist and service policy (b) Troubleshooting weak areas		(a) Company machinist on 24 hour call (b) Preventative maintenance program
1.5	Input volume		5 Takes		8 Takes		12 Takes		16 Takes		20 Takes
	Versatility to customer		(a) 10 fonts (b) Black only (c) Salesman correction		(a) 30 fonts (b) 1 to 2 colors and black (c) Proof on request		(a) 50 fonts (b) Black and color (c) 2 day salesman proof		(a) 70 fonts (b) Black and 4 color (c) 2 day proof		(a) 100 fonts (b) Black and 4 color press (c) 1 day proof

Personnel competence	(a) Basic make up (b) New in area (c) Apprentice	(a) Some make up (b) Some background (c) Puts time in	(a) Good make up (b) Moderate worker (c) General knowledge of process	(a) Accurate make up (b) Responsible worker (c) Working knowledge of process	(a) Fast accurate make up (b) Skilled worker (c) Operate computers (d) Minor decision maker
Status of personnel	Release 8 employees Retrain 2 employees Hire 5 employees	Release 5 employees Shift 5 employees Hire 8 employees	Keep all 10 employees within company	Keep all 10 employees within division	Release 3 employees Shift 2 employees Retrain 5 employees Hire 2 employees
Product quality	(a) Blotchy, broken type faces (b) Varying proportion (c) Broken crooked borders (d) Acceptable appearance	(a) Cracked or bubbled type (b) Uneven distribution of type and gray matter (c) Scum within border	(a) Mostly solid type faces (b) Good appearance (c) Eye-catching (d) Practical color break	(a) 95% solid type body (b) Clean, without scum (c) Color contract (d) Good alignment on type	(a) Solid type faces (b) Proportion exact (c) Color balance (d) Appealing appearance

Figure 12-4 Systems utility function.

*Adapted from Tharrington

customer and project quality are relevant aspects of the system objective, desirability of product to users. The system objective of life cycle cost is developed through the criteria of capital investment and page cost, and so forth.

Once a set of criteria has been determined, it is then necessary to establish what the various levels of performance (i.e., poor, normal, exceptional) are on each criterion during the planning horizon. To accomplish this task, the analyst talked to many salesmen who represented companies which had developed machines for setting newspaper type; he studied the ramifications of his own company's present method of setting type; and he also talked to other suburban newspaper printing organizations. From this analysis, the systems utility function and more specifically the measures of performance in quantitative and qualitative terms were developed (see Fig. 12-4).

To interpret this chart, consider for example the criterion of capital investment. Assume this criterion is defined as the total price required to purchase the machine for setting newspaper type and to operate it at an assumed rate for five years, less the salvage value at the end of the planning horizon. With this definition, the normal capital investment for the typesetting process for this size of suburban newspaper company would be expected to be around $550,000. The least costly methods would average about $300,000 and the most expensive methods would run about $800,000.

Since advertising is a major source of revenue to newspapers, the criterion of versatility to the customer was established. For typesetting machines in the $300,000 to $800,000 range, a versatility which includes 50 fonts, black and color print availability, and two-day salesman proofs can be normally expected. Exceptional versatility would include 100 fonts, black and four-color processes, and one-day proofs.

As an additional example of the systems utility function, consider this one developed by Ron White for determining the "Best Way To Raise a Child." This is a particularly difficult topic, because of problems in defining the performance levels for complex, qualitative criteria.

Note in this case, to establish the performance yardsticks for the various criteria that he and his wife felt were important, Ron needed to develop a set of comparables. These alternative child-raising styles would be derived from society in general, but more specifically from other couples who are of similar circumstances (i.e., age, class, region of the country).

By observing them, reading popular books and academic textbooks on how to raise children, Ron developed what he considered to be exceptional, average, and barely acceptable performance for each of the criteria. Since the criteria are basically qualitative in nature, it makes sense to describe them in words as opposed to numbers representing technical measures.

In the case of the criterion of attitude toward sharing, Ron defined exceptional performance as the child being generally glad to share and doesn't always have to be asked. If the child generally shares out of necessity or because it is required for good manners, that is average performance. Barely acceptable performance is when a child only shares when asked and argues against doing so.

Several points are important in this example. First, the comparable set should be for children of similar ages. That is, what is reasonable to require of a child at 18 years of age, can be very unreasonable at age 8. Secondly, the "exceptional" classification is never meant to be perfect. A child's performance can always be improved upon. Rather, exceptional is classified as that performance of the best 10% of the children in this group. Lastly, because the measurements are expressed in words they are subject to wider latitude of interpretation by different observers than would a more quantitative measure such as number of spankings per year. While number of spankings per year is relatively straightforward to measure accurately, that doesn't imply it is a better quide to performance than a more subjective measure. The test is to try to measure the relative value of something in a way which captures the basic nature of the underlying variable. If it is quantitative, we use numbers; if it is subjective, then we should use words.

SYSTEMS SIMULATION CHART

As was shown in the first part of chapter 7, to make an evaluation of the alternative system designs, we need to determine both what the system needs or desires and what the system outcome will be if we choose particular solutions. We have covered the systems needs and desires through the concept of performance measures and shown how this relates to the systems utility function. We now need to see how to determine the systems outcome.

We indicated in Fig. 7-4 that the systems outcome is defined in terms of What is the system going to get? and What is this to cost? Now that we have covered the concepts of system objectives, criteria, and performance measures, we can see that the question What is the system going to get? signifies performance which is beneficial or desirable. On the other hand, What is this to cost? refers to measures which are costly or undesirable. Here performance would be measured on the appropriate technical, economic, psychological, and political criteria.

Per se, we can't really answer this question until after an alternative is actually implemented. But that is too late and clearly not helpful to the decision process. We need to decide prior to implementation which alternative appears to be the best decision. Because we are

Rating

Weight	Criteria	0	1 Barely Acceptable	2	3 Below Average	4	5 Average	6	7 Above Average	8	9 Exceptional
8	Stress on Morals		Self-indulgent, disregards morals		Does as people with him do		Generally does right in crises		Seldom falters, feels sorry		Few mistakes, self-correction
6	Respect for Parents		Disconcern for parents' views		Doesn't stand by parents, some regard for them		Honors parents, but not always defending them		Consistently stands up for parents		Requests views, firmly defends parents
	Responsibility		Makes excuses, poor performance		Reckless approach to regular jobs		Performs required work acceptably		Does work well, willing to do more when asked		Does work well, does more on his own
	Attitude toward Parents' Word		Refusal to comply until forced		Begrudging compliance (back talk, etc.)		Compliance with limited unacceptance		Quick compliance, arguing the point to himself		Same, but attempts to dispel differences
5	Love for other Family members		Little regard for others' wants and needs		Convenient care for others' wants and needs		Genuine care, makes it known when he's put out		Genuine care, sacrifices when he must		Sacrifices, no 'martyr' image
3	Respect for Authority		Little concern for any authorities		Shows concern when authority present		Respect at most any time		Genuine respect and conformity		Same, attempts to learn also
	Ability to Adapt		Complains about any problem, uses any language, tone		Complains about any problem, whines		Similar, but only on more significant matters		Keeps his composure when arguing		Same, uses objectivity when discussing
	Attitude toward Sharing		Only when asked and argues		Out of necessity but resentful		Out of necessity or good manners		Glad to share when asked		Same, need not be asked
	Equality and Individuality		Child feels inferior to brother or sister		Child feels less loved than brother or sister		Child feels equal but parents have created competition		Same, but no competition		Same, and child feels his individuality
1	Wasting of Resources		Child consistently takes more food, material, etc.		Same, but not as often		Same, the child feels apologetic when occurring		Very seldom ever takes too much or wastes		Same, also he watches his time as well

Figure 12-5 Systems utility function.

*Adapted from White

216

dealing with a future time period, there are few, if any, facts. Mainly there is conjecture, supposition, and guessing.

To get a really good feel for this dilemma, consider one of the most important decisions in one's life, "selecting a wife" or "selecting a husband." How does one know prior to marrying a person how married life with that person will actually be?

What are the critical questions to ask (i.e., criteria)? How do you get information on the future "systems" performance? Basically, you are forced to make a simulation of the eventual outcomes.

Simulations are used to estimate the expected performance of a system, using a model of the actual situation.

When the United States was preparing to put a man on the moon, scientists didn't simply launch the manned space capsule to see how it would turn out. Rather, there was a very elaborate simulation process for determining what the outcomes would be if certain procedures were used. The problem was that we didn't know how people would react to a sustained period of zero gravity because man had never experienced that before. So NASA, the U.S. space agency, built a special apparatus to suspend astronauts in a swimming pool and try as best they could to simulate zero gravity conditions.

For airliner pilots, there is a very sophisticated computer-based simulation of flying conditions. The pilot sits in a cockpit which has all the same instrumentation, feel of the controls, sounds of the aircraft, visual pattern of various runways, etc. The airlines have found this is a very cost-effective way of testing their pilots in emergency conditions. But while very effective, it must always be remembered it is a simulated situation, not the reality. Nevertheless, what used to be done was to take pilots into the air and give them emergency aircraft situations. This was much closer to reality, but not very cost-effective. Moreover, many airplanes were damaged and many pilots were killed in using this approach.

Both the space and the airliner examples are very expensive and sophisticated simulations. We can however apply these concepts to much simpler examples. Sometimes, for a variety of reasons, either we can't or may not want to deal with the reality directly. So we need to simulate it with some type of model. As we saw in chapter 9, a model is any abstract representation of reality. The model can be very elaborate or alternatively very simple.

For our purposes, we will use four categories of models for determining expected performance of a system: (1) logic, (2) tests, (3) experience of other systems, and (4) on-the-job.

When one tries to reason through what would happen if he did such and such, we refer to this abstract method as using logic to simulate behavior of a system. Frequently we can perform one or more kinds of tests. This would be the case for example with the Environmental Protection Agency using a dynometer to estimate future gas mileage for a car, as opposed to actually driving it.

A third approach to simulation is to observe other comparable systems which are using the particular alternatives under investigation. That is, if you are thinking about buying a particular car, you could benefit from knowing what kind of performance other people have gotten using the same model under similar driving conditions. The last approach is to actually use the system for a test period in the actual situation. For example in the computer field, some companies run parallel tests of the computerized system and their manual method to develop more accurate estimates of future system performance.

To show how these concepts can be used, we will give examples of a system simulation for estimating mpg, determining the challenge of work, and the capital investment required for a typesetting process.

Systems Simulation Examples

In format, the system simulation chart is made up of the following parts: criteria, alternatives, expected performance, confidence, and support. For each of the feasible alternatives in the final set, the analyst needs to develop an estimate of what the expected performance would be over the planning horizon if that option were actually implemented. Further, the analyst needs to show what evidence or support has been used as a basis for drawing conclusions and to provide an overall confidence indication of the accuracy of the expected performance estimate.

As an example of a systems simulation chart, see Fig. 12-6. The criterion is gas mileage. There would be a separate chart for each criterion—cost, status, reliability, etc. The final set of feasible alternatives are the new 1982 Toyota Corolla, 1982 Honda Civic, and 1982 Plymouth Reliant. The 1981 Fort Pinto is new, but a year-older model. The 1976 Chevy Impala is the existing system modified through an engine overhaul and a paint job.

The analyst's best estimate of mpg that the various alternatives would get when the CDM uses the car in the Jones family systems context over the planning horizon of 1982-1987 is: Toyota—33 mpg, Honda—44 mpg, etc. The reality is that nobody knows what the Joneses will get in gas mileage during 1982-1987. But because gas milage is a relatively stable kind of variable, we should be able to get some good estimates of expected performance, but always less than 100% guaranteed as "fact."

The analyst feels the Jones family will get 33 mpg with the

Criterion	Alternative	Expected Performance	Confidence	Support
	1982 Toyota Corolla	33 mpg	LC	Inspected the sales brochure (3)* at the Toyota dealership, noted estimates by manufacturer. A friend of yours (2) indicates he gets good gas mileage with his 1978 Toyota.
	1982 Honda Civic	44 mpg	VC	Consumer Union report of February 1982 (7) ran extensive tests among many similarly equipped Hondas across highway and city driving. Environmental Protection Agency (15) rating for this car as shown on the window sticker are in close agreement. Hot Rod Magazine for October 1981 (4) reports essentially same gas mileage results.
GAS MILEAGE	1976 Chevy Impala	18 mpg	C	Have kept close tabs on the existing system for over four years (6). Engine overhaul should improve mpg somewhat.
	1982 Plymouth Reliant	21 mpg	NC	The new car salesman (18) at the Plymouth dealership indicated that the car gets good gas mileage. You got a much lower estimate using the gas gauge while driving a demonstrator around town (1).
	1981 Ford Pinto	26 mpg	C	Motor Trend report of October 1980 (8) ran tests over several traffic conditions using a 1980 model equipped with automatic transmission. Their findings are in close agreement with your personal driving comparisons (10).

Figure 12-6 Systems simulation chart: selecting a car.

*Numbers indicated reference source.

Toyota. The support for this conclusion comes from a sales brochure and a friend's driving experience with an older Toyota. This, while the best she could get within the time allocated to the study, only warrants a "Little Confidence" indication. Recall the discussion on confidence indications in chapter 11; The 1976 Chevy Impala is expected to get 18 mpg, if an engine overhaul were performed. The analyst is "Confident" since the CDM has worked with the existing system solution for four years. However, they are not "Very Confident" since the overhaul should improve the mpg but it is not known how much. In turn, the Joneses are contemplating moving to a new house during the planning horizon. This will require Mr. Jones to do more freeway driving, so the car's mpg would improve in comparison to the driving conditions of the last four years.

To see how these concepts would be applied in a more complex situation, consider the systems simulation chart of Fig. 12-7. Here the analyst is trying to determine what is the best accounting career field for the CDM. The alternatives are highly different in scope and the criteria are basically qualitative in nature. In studying this chart, remember the analyst is not trying to see what the situation is today, but rather is trying to estimate what it will be like during the planning horizon. In this case, tomorrow is estimated by using today as a basis.

The last example of a systems simulation chart was developed for the typesetting process example we have been studying (see Fig. 12-8). The alternatives in the final feasible set are cold metal process, computer typeset RCA, computer typeset IBM, hot metal process A and B. After studying the situation to varying degrees of detail, the analyst's best estimate is that if the computer typeset IBM method were to be implemented, it would require a capital investment of $476,800. The analyst arrived at this number from the purchase price quoted by the salesman and from checking with present users of this method in other publishing companies. The analyst's overall confidence in his estimate is "Low Confidence." The (2), (1), and (3), etc., in the support column are references to more detailed explanations of the sources given in the back of the system study under *Information Sources*. This approach can take the place of the more usual method of footnoting and bibliographies.

EVALUATION MATRIX

The purpose of the evaluation matrix is to give the decision-makers and analysts a means of establishing the relative ranking of alternative courses of actions by evaluating each system design and the explicit tradeoffs and risks involved with each choice. More explicitly, this is accomplished by making a complex, multidimensional compari-

Criterion	Alternative	Expected Performance	Confidence	Support
	Small C.P.A. firm	Constant interrelating among tax, audit, and management service problems for several different organizations.	VC	I have worked in this position at the existing system for over two years. (9) No major changes in functions are anticipated.
	"Big 8" accounting firm	Work is interesting and the clients are large enough to require sophisticated analysis. However, most staff are required to work exclusively on audit problems with little contact with tax or management services.	C	There was wide disagreement among Bill George, audit chief (7) and Sam Walker, UCLA intern (3), in this regard. Professor Brown of Cal Poly felt strongly that auditing was where most all college graduates started.
CHALLENGE OF WORK	Federal government agency	Work in both investigative and audit work with about 15% of time spent "undercover." Assignments vary and audits are conducted on an as required basis and may be with any civilian contractor. Investigative work may take one all over the country.	LC	Didn't get to talk to any staff workers. Only information was from Bill Ames, Division Chief of Audit Branch (1), and assistant personnel recruiter (12) during a campus interview.
	Medium-size manufacturing company	Work would be almost all internal to firm. Responsibilities would include receivables, payables, and costing. Some interaction with auditors and with company engineers. As assistant controller, I wouldn't take charge of major items for some time.	C	I have extensively studied the internal organization of this firm (2). My conclusions were further supported by extensive talks I had with Ms. Smith, controller (9) and Tom Green, vice-president of the firm (13).

Figure 12-7 Systems simulation chart: accounting career.

Criteria	Alternatives	Expected Outcome	Confidence	Support
	Cold metal process	$285,000	VC	This has been the existing system for four years. Machines fully paid off, no purchase required. Operating costs expected to increase by 5% per year due to age, company controller (7). Similar operating cost experience at *Sun Post*, controller (5)
CAPITAL INVESTMENT*	Computer typeset RCA	$410,000	C	Initial purchase price has been specified in contract by Jones RCA (9). Operating expenses have been verified by present users of equipment, *Pomona Journal* (14), *Chino Blade* (10), and *Oxnard Tribune* (22)
	Computer typeset IBM	$476,000	LC	Initial purchase price stated by salesman Clark, but he was unwilling to put on contract without formal approval of his boss (2). Present users experience 0 to 5% higher operating expenses than quoted (1), (3).
	Hot metal process A	$303,000	NC	Only have sales brochure figure (6). Unable to get names of references. Seems like a very low figure.
	Hot metal process B	$358,000	C	Purchase price specified in contract (10). Equipment has been around 5 to 10 years. Riverside Press says figure within 10% of their experience (19).

Figure 12-8 Systems simulation chart: typesetting systems.

*Capital Investment is defined as initial purchase price plus 5 years annual operating cost less salvage value.

son of the expected future performance of each alternative with the desired future systems performance.

The totality of information is displayed and summarized in a chart like Fig. 12-9. At the top are listed the *Feasible Alternatives* which are being evaluated in the system study. Each alternative is a workable design which will solve the stated problem. At the left of the chart are the *Criteria*. These specific dimensions of the overall system objectives are used to establish the relative worth of the various alternatives. In the study of any complex system, there is a need to develop multiple criteria that are multidimensional in basis. Then the criteria should be grouped and weighted according to their perceived degree of influence in substantiating the worth of all system designs.

The internal entries of the matrix consist of four elements: R for relative rating, C for confidence, U for systems utility and D for discounted utility:

Relative rating is the resultant comparison by the analyst of his best estimate of what the systems performance would be if the particular alternative under consideration were implemented (i.e., systems simulation chart), when compared to the level of performance that could normally be expected (i.e., systems utility function).

System utility is a measure of the contribution to the total utility of the system with performance on a particular dimension, if a specific alternative were implemented. It is determined by multiplying the rating times the relative importance of the particular criteria.

Confidence is a subjective judgment by the analyst of the probable accuracy of a particular system utility score reflecting the true (actual) utility to the system over the planning horizon. The four levels of accuracy are Very Confident, Confident, Little Confidence, and No Confidence. The appropriate level of confidence for each entry in turn is determined by the validity of the facts, estimates, opinions, experiments, etc., the analyst was able to gather to validate the various weighted scores.

Discounted utility is an explicit means of considering the accuracy behind each of the ratings of each alternative design. The subjective probabilities for each rating are multiplied by the respective systems utility score to give the discounted utility score.

Feasible Alternative

Criteria	Cold Metal Process Existing System				Computer Typeset RCA				Computer Typeset IBM				Hot Metal Process A				Hot Metal Process B			
	R	C	U	D	R	C	U	D	R	C	U	D	R	C	U	D	R	C	U	D
Very Important (Weight–5)																				
Capital investment	10	VC	50	45.0	7	C	35	21.0	6	LC	30	9.0	9	NC	45	4.5	8	C	40	24.0
Time element	10	C	50	30.0	1	VC	5	4.5	1	VC	5	4.5	10	C	50	30.0	10	VC	50	45.0
Important (Weight–3)																				
Page cost	1	VC	3	2.7	10	VC	30	27.0	10	VC	30	27.0	8	LC	24	7.2	9	LC	27	8.1
Maintenance of system	7	VC	21	18.9	5	VC	15	13.5	7	VC	21	18.9	7	VC	21	18.9	7	VC	21	18.9
Moderate Importance (Weight–1.5)																				
Input volume	0	C	0	0.0	10	VC	15	13.5	10	VC	15	13.5	6	C	9	5.4	2	NC	3	0.3
Versatility to customer	4	VC	6	5.4	8	C	12	7.2	10	VC	15	13.5	6	LC	9	2.7	6	LC	9	2.7
Personnel competence	6	C	9	5.4	2	C	3	1.8	4	LC	6	1.8	4	LC	6	1.8	6	LC	9	2.7
Minor Importance (Weight–1)																				
Status of personnel	7	LC	7	2.1	3	C	3	1.8	1	VC	1	0.9	5	LC	5	1.5	5	LC	5	1.5
Product quality	5	VC	5	4.5	10	VC	10	9.0	10	VC	10	9.0	7	NC	7	0.7	9	VC	9	8.1
Total Value	151				128				133				***176***				173			
Discounted Value	**114.0**				99.3				98.1				72.7				111.3			
Overall Confidence	76%				78%				74%				41%				64%			

R = Relative rating   C = Confidence   U = System utility   D = Discounted utility   VC = .9   C = .6   LC = .3   NC = .1

Figure 12-9 Evaluation matrix.

At the bottom of the evaluation matrix are the entries for total value, discounted value, and overall confidence.

Total value is an overall estimation of the relative utility or value to the system, if a particular alternative system design were implemented. It is. determined by adding the scores under the U column or system utility score for each alternative.

Discounted value is an overall indication of the utility or value to the system, considering the accuracy of the information gathered. It is given by adding the discounted utility scores for each alternative under the respective D column.

Overall confidence is a means of showing what the accuracy is overall in rating each alternative. It is determined by dividing the discounted value by the total value for each alternative.

Let's consider the completed evaluation matrix of Fig. 12-9 in more detail. At the completion of the analysis phase of the system study to determine the best means of setting type for a small, suburban newspaper, it has been established that there are five alternative designs which are workable. These are (1) a cold metal process, which is a slightly modified version of the existing system, and two revolutionary means of computerized typesetting referred to as (2) computer typeset RCA and (3) computer typeset IBM. The remaining two alternatives represent new evolutionary methods of setting type which while still manual, are considered by the industry to be very worthwhile advancements. These alternatives are designated as (4) hot metal process A and (5) hot metal process B.

In consultation with the CDM, the problem-solving group, and with workers, it has been established that there are nine criteria which reflect how the typesetting process can best relate to what the overall organization is trying to accomplish. The two most important criteria are capital investment and time element; they are given a relative importance weight of 5. Page cost and maintenance of the system are also important aspects of the system solution, though not quite as important as capital investment and time element. The least important criteria, status of personnel and product quality, have a relative importance weight of 1.

In general then, the problem-solving group feels that they would be willing to give up much in product quality to save some money on the capital investment. Additionally, if need be, they would also be willing to displace some workers to get a typesetting process which can print the newspaper in a quicker manner. Do you agree with this analysis? Should an organization have so little concern for the quality of its product or its employees? How is your thinking affected by the systems context?

Let's now look at relative ratings by considering the criterion of page cost. By looking at the evaluation matrix, it is clear that RCA, IBM, and processes A and B are all much less expensive than the existing system. Additionally, the computerized alternatives are as cheap as one could reasonably hope for. More specifically, look at the system utility function to see how much money it will cost to establish each page. For the RCA and IBM alternatives it will cost around $2.20 per page. Hot metal process B costs about $2.25 and the existing method costs a very expensive $2.95 per page.

In actual practice, these relative ratings are established in a reverse manner to the way we just interpreted them. First, the analyst determines the systems utility function by establishing the desirability of various levels of performance or costs. In this case, $3.00 per page is barely acceptable, $2.43 is considered average, and $2.20 is very inexpensive.

The analyst now gathers information about each of the five feasible alternatives, and then he estimates what the page cost will be in the systems simulation chart. Assume that for the existing system, he feels $2.95 is the best indication of page cost. By itself, one cannot say $2.95 is good or bad, or expensive or inexpensive. Only after comparing it to something can we make the statement that it is very good or very poor. After comparing the $2.95 to the costs shown in the system utility function, we can see the existing system is very expensive in terms of page cost; in fact, it is given a barely acceptable rating of 1.

The utility scores reflect the value of both the relative rating and the importance grouping. For example, each of the computer typeset methods scores an exceptional ten points on both page cost and input volume. However, since page cost is considered twice as important (3 vs. 1.5 weight) as input volume, the resulting utility score is 30 points versus 15 points.

The confidence indications show a summary judgment by the analyst as to how sure he feels he is about each rating. In general, the analyst feels confident to very confident as to rating the alternatives on the criterion of time element. This is so because he was able to get verified benchmark testing for each of the typesetting methods as reported in the Printer's Trade Association Annual report. However, regarding the status of personnel criterion he feels much less confidence with respect to all five alternatives. This makes sense, because of the uncertainty surrounding union demands and the lack of knowledge about how much the affected workers will be able and willing to learn new technical skills and social relationships. This overall confidence and support for the conclusions is given in the systems simulation chart.

By assuming confidence level values of VC = .9, C = .6, LC = .3, and NC = .1 (see chapter 11), the discounted scores for versatility to customer can be checked. For the existing system the systems utility score of 6 is multiplied by .9 for a VC confidence level. This results in a discounted utility score of 5.4. Similarly, the low confidence shown for rating hot metal process B on the same criteria results in a 2.7 discounted score (i.e., 9 x .3).

Hot metal process A has a total value of 176. This is derived by adding the U column. More specifically, 45 + 50 + 24 + 21 + 9 + 9 + 6 + 5 + 7. Further, process A has the highest total value of 176, when compared with values of 173 for process B, 151 for the existing system, down to the least total value of 128 for the RCA alternative.

The discounted value is determined by adding the D column for each alternative. For the cold metal process design, the discounted value of 114.0 is derived by adding 45.0, 30.0, 2.7, 18.9, 0.0, 5.4, 5.4, 2.1, and 4.5. Interesting to note, the existing system has the highest discounted value whereas it was only third highest in total value.

Overall confidence for hot metal process B, for example, is determined by dividing the discounted value 111.3 by the respective total value 173. This results in a percentage of approximately 64. In comparison with the other alternatives, it can be seen that the supporting information for this alternative is not as strong as for the existing system or the computerized methods, but much higher than hot metal process A. In particular, we note that the overall confidence indication is just a summary device of how confident the analyst is of his ratings. It's importance is secondary, since it is different from the worth or value of the alternatives.

RATIONAL ANALYSIS OF ALTERNATIVE RANKINGS

Thus far we have covered what each of the numbers and symbols means in the evaluation matrix. The most important phase of the systems approach comes next—that of interpretation of these results, to aid in the selection of the best alternative for this particular system. Our decision will be reached by evaluating total scores and variation among alternatives, weighing of risks, considering where the search for more information can be helpful, and doing an impact analysis.

Total and Discounted Value

It has been stressed many times that the evaluation matrix should be used as an aid to the selection process. One should not pick the alternative to implement strictly on the basis of the highest total value. On the other hand, the totals reflected in the matrix do give a summary overview of what the analyst has determined.

In the case of the typesetting system, hot metal process A and B have a significantly higher total value than either computerized method (176 and 173 versus 128 and 133) and a somewhat greater value than the existing system (176 and 173 versus 151).

When the discounted value is considered, a different perspective is gained. In this situation, cold metal process and the hot metal B have the highest values of 114.0 and 111.3 respectively, whereas the computer typeset methods of RCA and IBM have middle values of 99.3 and 98.1 respectively. Clearly, the lowest discounted value of 72.7 belongs to hot metal process A.

An interpretation of total and discounted values would go more or less as follows: If everything should turn out like the analyst's best estimates have indicated, then hot metal processes A and B are clearly the best. No further consideration need be given to either computerized method. The existing system should not be ruled out yet, since while it is not the highest valued, it is relatively close. Comparison between process A and B would have to involve factors not considered in the evaluation matrix. These would include criteria not significant enough to be entered on the chart, requiring further reference to the likelihood of the alternative to be implemented, etc.

But the above discussion assumes the information is all valid. We know this would not be true in the analysis of the future performance of any complex system.

Therefore, analysis should center on what the discounted value indicates. Since the computerized methods score relatively low here also, I would eliminate them from further consideration. Hot metal B and the existing system score the best when the information is discounted for probable inaccuracy. Hot metal A does very poorly. Evidently, very little is known about this process as indicated by an overall confidence rating of 41 percent. However, since it scored so well on the total value, we would keep it in the final analysis.

Impact Analysis

When we initially decided on the scope of our study and defined the system and its environment, we realized that we could not handle all the possible combinations of future environmental factors. Therefore, we opted for the idea of scenarios. This entailed estimating the most likely situation for three or four major environment factors. We then went ahead and developed different system designs and rated their expected performance in this assumed environment.

The purpose of impact analysis is to see what the effect would be of varying the scenario, on the *desirability* of the alternatives. That is, if the environment turned out to be different from what we had anticipated, would that affect our choice of which alternative is best?

For the typesetting example, let's consider one factor in the scenario such as the amount of competition we face from other newspapers in terms of circulation, advertisement, etc. In looking at the future planning horizon of five years, we assumed that the most likely development would be a slight increase in competition from the *Los Angeles Times.* This newspaper has hinted several times that it was thinking about putting out a special edition aimed at our community. However, since it takes several years to establish this type of enterprise, we feel little immediate threat.

Our optimistic estimate is for the *Los Angeles Times* not to enter the market at all, or if it did, it would fail. Thus, this condition assumes things will stay the same as the present, which implies essentially no competition.

The pessimistic estimate is for the *Los Angeles Times* to enter this market in a major way. This condition assumes a significant increase in competition from a major newspaper.

The question that now needs to be addressed is, How will this affect our choice of alternatives, which has been based on the most likely estimate of just a slight increase in competition?

In talking with the CDM and others, it has been stated that if the pessimistic condition were to occur, our newspaper would have to significantly raise its level of competitiveness through increased advertising and circulation. This in turn implies greater versatility to the advertising customer, and bringing the product quality of the newspaper itself up to the much higher level of the *Los Angeles Times.*

Relatively speaking, then, the importance of the criteria of versatility to customer and product quality would increase significantly. In both of these cases, hot metal processes A and B are better than the existing system and thus would become even more desirable as the best choice. Between process A and B, both are equally versatile to customer so there is no change. However, in regard to the quality of product, process B can do a better job which thus increases its value as an alternative.

Under the optimistic assumption, implying no competition, versatility to customer and product quality would have slightly less importance than at present. However, this would have *no significant impact* on the decision, since these criteria already have the lowest weights and the point values are such that no change in alternative rankings will occur.

Gathering More Information

At this point in the analysis, the computerized methods have been eliminated from further consideration. Hot metal process B has been shown to be a better choice than the existing system because it has essentially the same discounted value and the potential for a much higher total value. Additionally, considering impact analysis increases the desirability of process B.

It is clear by the confidence indications that all information has not been gathered and therefore, we are not 100 percent sure that we have the "correct" answer. But that is not the appropriate question to ask. Our goal is not simply establishing the truth; i.e., determining what is. What we are interested in is making a good decision. Collecting information is time-consuming and costly in terms of dollars, manpower, frustration, energy, etc. Therefore we need to ask, Is there any more information we should gather which will increase the quality of our decision enough to offset the additional cost of gathering it? And if so, where might our effort be best spent?

We could reconsider the criteria and their relative weights. We could also increase the confidence levels of any alternative, by gaining more evidence to support our ratings. The ratings themselves could be changed as a result of further investigation. These considerations then need to be weighed against how much it costs to get this increased knowledge and what difference might it make vis-a-vis the decision on the best alternative.

For example, the confidence levels on the computerized methods, while high, could be increased through more information searching. But why do it? This will not make either of them the best alternative. However, if more information were gathered on Hot metal process A on the capital investment and page cost criteria, this could make a significant difference; i.e., it could change the decision on the best alternative. Increasing just the confidence levels on these two criteria to "Very Confident" would raise its discounted value by about 40 points. Then process A would have a higher score on both total value and discounted value than process B. On the other hand, how costly is it to get that information? Since they would still be essentially of the same value, why not just pick process B now?

What has been discussed in this section is the "rational" use of the information summarized in the evaluation matrix. We will need to discuss the concepts of The Decision Process—chapter 13, Implementing the System Solution—chapter 14, and Communicating Study Results—chapter 15, before we can say which alternative should really be selected as "best."

SUMMARY

In this chapter, we have given an explanation of the preference chart, systems utility function, systems simulation chart, and finally the evaluation matrix. These tools are used to aid the analyst and decision-maker in the evaluation process.

We now have a way which summarizes explicitly for decision-making purposes what has been accomplished, evaluates how good it is, and identifies areas where further study is warranted. This results in a very effective communication between the analyst, the chief decision-maker and other interested parties.

The evaluation matrix offers a very practical way of handling problems which have multiple objectives of both a qualitative and quantitative nature. In addition each objective can be assigned different levels of importance in relationship to the final decision. In effective decision-making there is need to know how valid and reliable is the information used to rate the worth of various alternatives and how crucial these various factors are to the final decision. These considerations are made explicit.

With the systematic systems approach, the decision-maker can determine the best alternative under study using multi-dimensional objectives and explicit considerations of the validity, reliability, and sensitivity of the model results to any final recommended decision.

Class Exercise 12-1

SELECTING A MATE

The purpose of this classroom exercise is to establish what are the most important characteristics in selecting a mate in marriage, to live a happy and productive life together in the world of today and tomorrow (i.e., 1980-2000 time frame).

All questions are to be answered as a *group consensus.*

1. Establish exactly *9* criteria which your group feels are the most important facets of a mate. Does it make any difference whether the potential mate is a woman or a man?

2. Rank in order the criteria developed in question 1 from most important to minor importance. State your means, as a group, for handling disagreement of importance levels among the members.

3. Develop workable definitions for explaining what each of the three most important criteria means (one or two lines should be sufficient).

4. In a survey of some tens of thousands of divorced couples, Ann Landers

determined that the three most important factors leading to divorce were considerations of (a) money, (b) sex, and (c) in-laws. How does this information relate to what was developed in question 1? How reliable are these survey results?

5. Given that you have developed some useful criteria for a marriage situation, consider the following concepts in how you would determine, prior to marriage, how successful marrying a particular person would be.

Through history, marriages have been formulated by methods such as:

(a) parents got together for reasons of territory and decided that their respective children should be married;

(b) in the 1800s settlers in the western United States selected a bride out of a mail-order catalog;

(c) in the early 1900s, a young man and young woman dated with the supervision of a chaperone;

(d) in the 1950s, a young man and woman dated without a chaperone;

(e) in the 1970s, a young man and woman lived together prior to marriage.

From an information processing standpoint (i.e., not considering moral values), consider the utility of each of these methods in selecting a good partner.

Class Exercise 12-2

AIRPLANE CRASH GAME*

It is approximately 10:00 A.M. in mid-July and you have just crash-landed in the Sonora Desert in the southwestern United States. The light twin-engine plane, containing the bodies of the pilot and co-pilot, has completely burned. Only the airframe remains. None of the rest of you have been injured.

The pilot was unable to notify anyone of your position before the crash. However, ground sightings, taken before you crashed, indicated that you were 65 miles off the course that was filed in your VFR flight plan (i.e., from Los Angeles to Tucson via Palm Springs and Yuma). The pilot had indicated before you crashed that you were approximately 70 miles south-southwest of a mining camp which is the nearest known habitation.

The immediate area is quite flat and except for occasional barrel and saguaros cacti appears to be rather barren. The last weather report indicated that the temperature would reach 110 degrees, which means that the temperature within a foot of the surface will hit 130 degrees. You are dressed in lightweight clothing—short-sleeved shirts, pants, socks, and street shoes.

*This is a slightly modified version of a game developed by the U.S. Air Force Survival School in Elgin, Florida.

A. What is the problem?

B. What are the major criteria you should use to evaluate potential solutions?

C. What is the best solution to select for this problem?

Before the plane caught fire, your group was able to salvage the items listed below.

D. Rank these 15 items within 3 groups (most important, important, and minor importance) according to how they would help you fulfill your group strategy. Put 5 items in each group.

_____Flashlight (4-battery size)	_____ Bottle of 1000 salt tablets
_____Jackknife	_____ 1 quart of water per person
_____Sectional air map of the area	_____ A book entitled *Edible Animals of the Desert*
_____Plastic raincoat (large size)	_____ A pair of sunglasses per person
_____Magnetic compass	_____ 2 quarts of 180 proof vodka
_____Compress kit with gauze	_____ 1 top coat per person
_____.45 caliber pistol (loaded)	_____ A cosmetic mirror
_____Parachute (red and white)	

E. How confident are you of your information in ranking these items? How can you evaluate the worth of these items?

F. The Air Force gives one correct answer to this problem. How do your results compare? How do we determine who is right?

Class Exercise 12-3

COMPUTER VENDOR ANALYSIS

In class exercise 11-3 we assumed that a company had sent out a RFP for purchasing a small business computer system. After evaluating the results, the company analyst has come up with the accompanying evaluation matrix of the three finalists.

1. Are these criteria the most important ones to be considered? What about the weighting? What are the implications of applications software having a weight of 4 and the computer hardware system weighting 1? Do you agree with this?

2. How would you go about establishing a rating for the worth of the various vendor's applications software? How could it be that there are so few 10's in the matrix?

Vendor Evaluation Matrix

Alternatives

Criteria		ABC				MNO				XYZ			
		R	C	U	D	R	C	U	D	R	C	U	D
Applications Software	4	5	LC	20	6	6	C	24	14.4	7	VC	28	
Risk		4	C	16	9.6	6	C	24	14.4	6	VC	24	
Price	2	2	VC	4	3.6	1	VC	2	1.8	10	VC	20	
Ease of Use		9	C	18	10.8	8	C	16	9.6	7	C	14	
Growth Capability		6	VC	6	5.4	8	LC	8	2.4	8	C	8	
Service	1	5	C	5	3.0	8	C	8	4.8	8	C	8	
Computer System		2	LC	2	.6	7	C	7	4.2	6	LC	6	
Total value			71				89				108		
Discount value				39				51.6					
Overall confidence				55%				58%					

3. For price, what considerations should be included? How does the planning horizon of five years influence cost considerations?

4. To which vendor does the analyst assign the highest confidence ratings? The manager has asked why the analyst isn't completely confident on all the ratings. What is your response?

5. Figure out the discount value for the XYZ proposal. Which vendor has the best total value? Which has the best discounted value? Who would you recommend? Why?

6. After hearing your recommendation, the CDM requests you to go collect more information to increase all confidence levels to very confident. He is especially concerned with the low confidence on the ABC proposal. What do you suggest?

Chapter Thirteen

The Decision Process-
Selecting
the Alternative

We have pointed out that the role of the analyst is to come up with the "best" solution to the systems problem. Up to this point, we have assumed that the "rational" approach we have taken is in fact the only reasonable method for improving systems effectiveness. Unfortunately, people-based systems are not this simple. Many times, pure reason doesn't assure that the solution with the best value as indicated in the evaluation matrix will always be selected for implementation.

This isn't all bad, since we know the analyst can't "prove" beyond a shadow of a doubt that the recommended solution is the "correct" solution. Further, because we are dealing with a future time period, the analyst's information base is by necessity less than 100 percent accurate. Therefore, we should be surprised and frankly disappointed if everyone accepted the study conclusions without discussion and without developing different conclusions based on their different perspectives. But this difference of opinions is not the major reason for system study results not being accepted. Rather it is a case of what is best for the system under study not always being best for the subsystems or the affected systems; this results in negotiation and produces compromises necessary to insure solution implementation.

To gain the background necessary to see how decisions are actually made, we will discuss the major factors which affect the decision process. Then we will see how these factors can be used in selecting the alternative to be implemented. Lastly, a business decision example will be given to amplify the major points developed in this chapter.

FACTORS AFFECTING THE DECISION PROCESS

In this section we will develop the background necessary to understand the major factors that affect the decision process. We will discuss the concepts of individual versus organizational player, goal congruence versus divergence, CDM versus individuals, reward versus penalize, Maslow's hierarchy, negotiation, and compromise and accommodation.

Individual versus Organization Player

Every person who is a member of any type of organization (i.e., business, church, university, family, etc.) is expected to play various roles. Being a minister, a mother, a wife, a son, a vice-president of marketing, a teacher, a hippie, a soldier, etc., are all examples of roles which make various demands on people to act in certain ways (Schein, 1970).

Organization Player: This is a person who plays the particular role perfectly. As an example of this concept, consider a person who is regional vice-president of marketing of a major cigarette company. This person, in accepting this job with this organization, is expected to do certain things. His role requires him to further the best interests of the organization through what influence he has as V.P. of marketing. The fact that he personally thinks cigarettes are harmful and should not be sold is irrelevant (Ever see a job description which said anything about personal ethics?). Further, his role requires that a system plan which changes the configuration of the regions, which gives the total organization a better competitive posture, should not be fought because it results in much less status for him personally. That is, his region is now smaller, therefore he has less influence and less salary. Only if the organization is not really better off should the V.P. of marketing oppose the plan.

What is being described is the perfect role player. The person who when he accepts a position with an organization, accepts the role of organizational player fully without thought of personal consequences.

Individual Player: This person is the exact opposite of the organizational player. The individual player is strictly interested in what affects his personal interests. The organizational position he holds is merely a device which he can use for personal gain. Be sure to note that the individual player is not necessarily all bad from the organization's point of view. He may have gotten himself into a role such that when he works very hard to further his best interests, he also furthers the organization's interests. Additionally, what is in the best interests of the organization can also be in his best interest. In these later cases, the individual player will act just like the organizational player.

Goal Congruence versus Goal Divergence

It is very important to realize that people in an organization are members of (at least) three major systems—organizational, group, and individual. The overall organization makes demands on an individual through his position with the organization. The immediate functional area or task group that a person belongs to is generally a very powerful influence on a person's behavior. Additionally, the individual himself is a system in which demands are levied by himself, his family, and other forces outside of the work organization.

Goal Congruence: This is a situation where the goals of the three major systems (i.e., organization, group, individual) a person belongs to while in an organization are completely in harmony.

Figure 13-1 shows a case of goal congruence where the demands on an individual by the organization, his task group, and his own interests point in the same direction. In this situation, whether a person is an organizational player or an individual player is not important because the required or desired action is the same.

As an example of this situation, consider a researcher who works in the R & D section of a company. If this person makes a brilliant discovery which allows the company to make a major product advance-

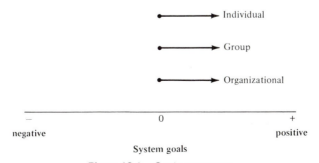

Figure 13-1 Goal congruence.

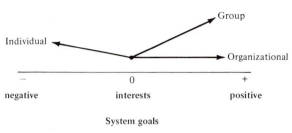

Figure 13-2 Goal divergence.

ment, then the company gains, the R & D group gains, and he gains personally. In this example, the researcher pursuing his own goals, even as an individual player, is also working in the best interest of the organization.

Goal Divergence: The situation displayed in Fig. 13-2, where the best interests of an organization and an individual are completely opposite is called goal divergence. Because the individual is not a perfect organizational player, but primarily interested in his own welfare, ready acceptance of this plan cannot be expected.

An example of goal divergence would be where the analyst has determined through a systems study that the organization as a whole would be best served by reducing the commission rate paid to the company salesmen. Assume it can be shown that a better benefit/cost ratio for the company can be gained by reducing the salesmen's commission from 10% to 6% of gross sales. While this will result in slightly less gross sales, it will yield greater net profit for the company. Assume further that the head of the marketing department's salary is determined as a base salary plus a percentage of sales.

Overall the company stands to increase its ROI from say 10% to 12% with a net increase in profits of $3.5 million to $4.0 million. The marketing section itself will become more cost-effective, because while it will bring in approximately 5% less sales it will do so by reducing selling costs by 15%.

When the analyst presents the study findings to the president of the company, how do you think the marketing V.P. will react to the study recommendations? (Assume the analysis is correct.) If we look at the results from the viewpoint of the marketing V.P., we see that (1) he will have to answer at least implicitly why he had not seen the need for this change; (2) he personally will bring home $10,000 less per year in salary (i.e., $40,000 now to $30,000) and (3) he is going to have to explain to his salesmen that even though they are personally going to lose money ($25,000 to $20,000 yearly commission), it will be better for the company overall.

If the marketing V.P. were a perfect organization player, his personal goals would not be a factor at all—the only consideration he would have is whether or not the organization as a whole gains. If it does, he will support it. The fact that he and his men lose out personally would not be a consideration. In this case, however, assume that the marketing V.P. is an individual player and sees that he is going to lose out personally if this plan is put into effect. Therefore, he will try to torpedo it subtly or possibly just through brute force.

CDM versus Individual

We assumed in our original definition of the chief decision-maker that he was the one who had full authority over the acceptance or rejection of the system study results. In subsequent chapters, however, we have given many reasons why the world of systems is not that simple. There are many people in the decision process such as staff people like lawyers, accountants, purchasing agents, etc., who can veto an alternative. By showing that it doesn't meet the requirements for a particular law, the alternative becomes infeasible and thus is rejected.

This type of veto power is reasonable and helpful. But what about the case where a line officer, who is an individual player, is so powerful that when he says the favored alternative is bad, the CDM won't resist or overrule him? Oftentimes, the CDM is not an individual but a committee with a chairman. What if the votes against a particularly good systems alternative are enough to overrule the chairman's power?

We see this latter type of situation frequently in political (governmental) systems where lobbying from special interest groups can cause the resulting decision of what is best from an overall systems standpoint to be modified significantly in favor of powerful subsystems. As we will point out later, these resultant modifications aren't always bad and often lead to a really better overall solution. But whether the result is good or bad, the consideration of the relative power between the system and subsystem representatives is a fact of life in all organizational systems (i.e., church, education, business, family, etc.), not just among political units.

Reward versus Penalize

There are basically two ways to get people or subsystems to go along with an idea. You can either reward or penalize the person or group. This is the old "carrot and stick" technique. The carrot approach says in effect that if you want to move people, you need to entice them by giving them something they want for going along a path you desire. On the other hand with the stick approach, if you want people to do something, you just keep penalizing them until they do it.

These thoughts are basic management styles usually called Theory Y for the carrot approach and Theory X for the stick approach. With the carrot or Theory Y approach, we give the person more of what he or she wants, as an inducement to go in the direction we desire, than the person would personally have otherwise. For example, if you do a good job for the company as management defines it, you get a promotion and/or a pay raise. When you got good grades at school, your parents probably rewarded you with money or the use of the family car or maybe just a smile while telling you they were proud of you.

With the stick or Theory X approach, we penalize a person if he doesn't go along the planned approach, such that he will be left worse off than he presently is. When you got bad grades at school, your parents might have punished you, took away some privileges such as using the family car, going out on weeknights, etc. In business this takes the form of losing your job, being passed over for promotion, not sharing in the company bonus, being sent to the company offices in Timbuktu. In the movie *The Godfather,* when one person didn't go along with the proposed plan, he found the head of his favorite horse in his bed.

The overall point is this, whether the reward or the penalty approach is best in motivating an individual player to go along with the organizational view is debatable. The reward approach is certainly more enjoyable for everybody concerned and most often is highly effective. In other circumstances, the penalty approach must be used to get effective results. From the analyst's point of view, he needs to assess the management style of the CDM to see which approach will be used in differing situations.

Maslow's Hierarchy

In the previous subsection, the carrot and stick approach to motivating people was discussed and several examples given. But what wasn't covered was the underlying principles of motivation.

These principles have been described by Maslow as based on human needs in his classic work, *Motivation and Personality.* He classified needs as physiological, safety, social, egoistic, and self-fulfillment.

> **Physiological needs** relates to one's requirements for living to include water, food, air, sleep, sex, shelter for protection from elements, handling disease, etc.

> **Safety needs** are concerned with protection against danger, threats, deprivation, etc.

Social needs relate to the desire for belonging, for association, for acceptance by one's fellows, friendship, love, etc.

Egoistic needs are concerned with self-esteem and reputation in the way of self-confidence, independence, achievement, competence, knowledge, status, recognition, etc.

Self-fulfullment needs are concerned with self-actualization through self-development, creativity, reaching one's potential, etc.

Maslow arranged these needs in a hierarchy with physiological being the most basic, then higher needs of safety, social, egoistic, and the highest needs being self-fulfillment. It was Maslow's contention that the lowest needs must be completely satisfied before one could move on to the next level of needs. That is, man lives for bread alone when there is no bread. But once he gets bread, he now thinks about his safety needs of keeping his job.

But we all know of cases where an artist's most important need is to create and he is willing to forego food to buy more paint. Or consider the people who knowingly went to their deaths by setting themselves on fire, as a protest against the Vietnam War.

Figure 13-3 is given as a modification of Maslow's theory which keeps the major classification of needs and the hierarchy. It shows that generally people want to satisfy most of their physiological needs first, then their safety needs, etc. But it also allows for the more realistic

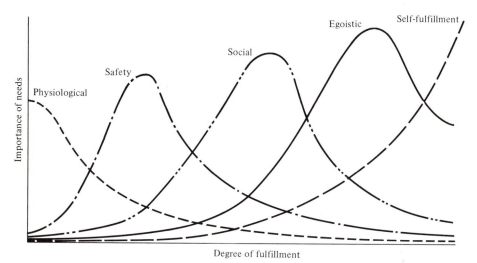

Figure 13-3 Goal congruence through roles.

situation where we as humans are trying with varying degrees of effort to work on all needs at the same time.

Finally, two points about the needs of humans are, first, a satisfied need is not a motivation of behavior. Only when you don't have it or it is threatened to be taken away is it a motivation factor. Secondly, the lower physiological and safety needs are relatively easily satisfied, particularly in the United States, and have fairly well-defined limits. However, the higher needs are not being satisfied as well in general, and even more importantly, egotism and self-fulfullment have no upper limit. That is, they can never be fully satisfied. Man always wants more!

Recall that in the book *The Agony and the Ecstacy* we told how Michelangelo had received the highest recognition for his sculpturing and painting. Yet, at the age of 89, on his deathbed, he prayed to God to spare him a little longer so that he could create one more work of art. If this wish had been granted and he had the chance to finish that work, he would at the age of 97, I have no doubt, asked God for time to create one more, etc. In other words, man is a wanting animal who can never be fully satisfied.

In summary, human beings generally act in ways which:

satisfy physical wants

bring security and release from worry and anxiety

lead to feeling loved and wanted

will gain achievement, recognition, and respect

will lead to self-development and creative challenge

Negotiation

We have talked so far about the hierarchy of human needs, but we haven't discussed how to determine what they are for a particular individual at a certain point in time. Here we are using the term *individual* generically to include the decision-maker who is representing some subsystem or a group acting as a CDM for an overall organization, etc.

Most people assume that to motivate people you need to give them more money. In many cases this is true and much of what we hear on TV and read in the newspapers is involved with salary disputes in which the various sides try to negotiate a financial settlement. There is much to be said for using money as a medium for enabling a person to choose their own needs to satisfy, as opposed to trying to figure out each person's needs.

The underlying concept of negotiation is to have two or more parties ask each other what their respective needs are. That is, to arrive at an agreement, each party needs to know what the other parties want out of the deal. In other words, what has to be "thrown in the pot" to make it a good deal for each group?

Thus each party has a much better feel for what it will take and in what form to get a various alternative solution accepted. As Nierenberg (1973) suggests, if this can be done in a joint spirit of cooperation among the parties, it can generally be very effective. It tends to cut down on the amount of gamesmanship usually seen, which often results in very disruptive meetings leading to divisive stalemates.

Compromise and Accommodation

In trying to get a solution accepted, the analyst is generally faced with this question: Is it better to press hard for the "best" solution from the organization's viewpoint which may be hard to get accepted because it is not in the direct interest of various subsystem groups, or is it better to seek a less effective solution from the organization's viewpoint which has an excellent chance of being accepted? In other words, when should one advocate compromise and accommodation? When is it beneficial to accept less than the best? When is second-best best?

This point is beautifully described below in a quotation of Walter Murphy from his book, *Elements of Judicial Strategy:*

> Supreme Court Justices have as much independence in the sense of job security as any free society can safely give its rulers. With this protection, a Justice might reason that he can, indeed must, vote his conscience and speak his mind on all cases which come before him regardless of the reaction of his colleagues, other government officials, interest-group leaders, or the public at large. The Constitution, the Justice could say, gives him a share of authority to decide cases according to the law as he sees it. What other people do with their share of law-making and policy-making authority is their business.

> For a Justice who reasons in this fashion, moral choice is clear and his conscience easy. Except for trivial incidents, he cannot compromise, he cannot accommodate, he cannot delay or retreat. Certainly one can understand this point of view, and while admiring it still wonder whether a Justice could not in good, though less easy, conscience think in a very different fashion. A Justice who perceived that his power was limited might feel that he should establish a hierarchy of values he wanted promoted. Every decision, a policy-oriented Justice would believe, requires the expenditure of such scarce resources as energy, time, prestige, or good will. Indiscriminate use of these resources could leave him or the Court exhausted at crucial moments. Thus, the Justice might decide that, since he cannot do all the good he would like, he must allocate his resources to accomplish those ends which he thinks most

important, knowing that this allocation may mean that other tasks will not be accomplished.

Where speaking out in favor of a given policy would be certain to antagonize his colleagues or to be voted down in conference, or, if pushed through the Court, to provoke bureaucratic resistance, hostile congressional legislation, executive refusal to enforce, or public rejection, the Justice might well ask himself whether he has the moral right to try to persuade his colleagues to adopt such a policy or to vote for it himself, even though he believes the policy involved to be morally correct, constitutionally impeccable, and, if accepted, conducive to the interests of the nation. Under such circumstances, he might decide that it would be more moral to compromise and aim for a lesser but possibly attainable good.

A Justice faces the same problem in opposing a policy which is sure to triumph in the Court or in the other processes of government. He may find that his choice is whether to fight and lose all or to compromise and accept a lesser evil. To select a lesser evil is a hard task, but one which practical politicians frequently face and moral theologians respect. Thomas Aquinas stated that "a wise law giver should suffer lesser transgressions that the greater may be avoided." As chapter 3 pointed out, a Justice may reasonably decide that to refrain from a searing dissenting opinion with an open plea to Congress to reverse a particular policy is a lesser evil, since such a dissent might so antagonize his colleagues on the bench as to cause him to lose influence over them in equally or more important decisions.

Indeed, a Justice might reverse the usual question and ask himself whether it would be moral for him to sacrifice making some immediate improvement for the sake of a future and greater gain. He could decide that he should make such a sacrifice if on the available evidence he believes that: (1) doing the smaller good now would materially weaken the chances for later accomplishment of the greater good; and (2) the ultimate advantage of the greater good would outweigh the disadvantages of waiting. In discounting the long-range goal, the Justice must keep in mind that he is dealing with human beings of limited life span. As Keynes is said to have remarked, "In the long run we're all dead."

Murphy's point is that even Supreme Court Justices who have great personal independence through freedom from external pressures, still find a need for compromise and accommodation to best accomplish effective system change.

What can a systems analyst learn from these ideas? He needs to understand that he cannot always demand the "best" organizational solution. To be effective, he must learn to pick his spots. He will have to learn when to push strongly for the "best" solution for system change and when to compromise and accommodate (i.e., when less than best is actually best).

HOW PEOPLE REALLY MAKE DECISIONS

With the background gained from the previous section on factors affecting the decision process, we are now ready to use the decision framework (Fig. 13-4) to see how people really make decisions. We also will see why this approach makes more sense than the strictly "rational" analysis.

The first question the analyst needs to ask is whether he should be a reporter or a salesman. If he decides he should be a reporter, then his task is relatively straightforward. In his presentation to the CDM and others, he stresses what he has determined and shows the relative ranking of the alternative solutions according to "Total Systems

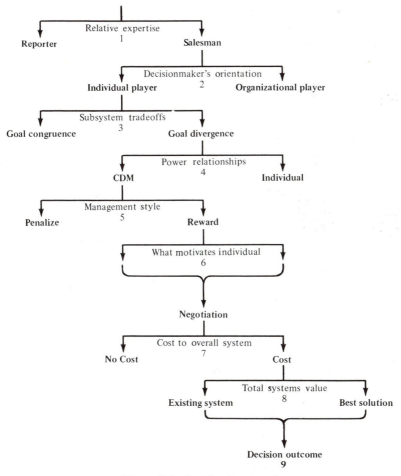

Figure 13-4 Decision framework.

Value." Who or what is affected is not his concern per se. He lets the "chips fall where they may."

We are assuming the analyst has the relative expertise (level 1, Fig. 13-4) in this situation, so that it is appropriate from an organizational standpoint that he be a salesman. This concept is discussed in detail in chapter 15, where we analyze the roles of reporter versus salesman.

Decision-maker's Orientation

If because of the circumstances, the analyst needs to be a salesman, then his first concern is knowing what is the orientation of the person or people he is going to need to sell the solution. In level 2 of Fig. 13-4, the major orientations are organization player vs. individual player. If the person that needs to be sold is an organizational player, then he is primarily concerned with what is best for the organization. Thus the analyst should stress the organization benefits of the various solutions. On the other hand, if the CDM is an individual player who is thus primarily concerned with the impact on himself of the various alternatives, then the analyst has to proceed to level 3.

Subsystem Tradeoffs

Whether there is goal congruence or goal divergence in this situation determines how the analyst should proceed. In goal congruence, what is best for the organization is also best for this individual. Therefore, the analyst should show what the overall benefits and costs are to the organization of the various solutions and show why the recommended alternative is best. This assumes the individual decision-maker will see that the resulting consequences of implementing the recommended alternative are also favorable to himself personally. If this is not the case, the analyst can, off the record, show the CDM how he will personally benefit from going along with the preferred system solution.

Power Relationships

If, in fact, there is goal divergence such that what is best for the organization is not best for the individual, then the analyst needs to estimate how important this difference actually is. In level 4, the basic power relationship is determined. If the system CDM has the power to decide what solutions will be implemented, then the analyst needs to look at the CDM's management style. If when there are differences, the CDM tends to penalize people who don't go along with the program, then the analyst should sell the recommended solution as what is best for the organization. The CDM will likely hint to the individual that if he doesn't go along, he will lose even more.

The analyst can suggest the recommended alternative and let the chips fall where they may. This solution recognizes that while the individual player is left with less than he presently has, this is the price that must be paid to get the best overall solution.

Management Style

The other approach, when the CDM has the power to make the system decision, is for his management style to be to share the overall system benefits with the individual left less well off under the recommended solution. This sharing also comes about in level 4 when it is determined that the CDM doesn't have the power to get the recommended alternative selected. This could be because the individual player can veto the plan or will torpedo the actual implementation or there is enough votes to override the CDM.

In the case of the other individual having this power, if he vetoes the recommended alternative, we are left with the existing system. Thus, the question becomes, should we share some of the system benefits which would be derived from implementing the recommended solution with this individual? In this way, while not getting the most benefits, the system will at least get more than it presently has with the existing system.

Thus, in our terms, the "buying off" of an individual who has the power to veto our plans, and the rewarding of an individual with more benefits than can be expected when the CDM has the power, results in a sharing of the system benefits.

What Motivates the Individual?

To make this choice effectively, the analyst needs to know what motivates the individual. Earlier we described human motivation in terms of Maslow's hierarchy. An individual, and many also say organizations themselves, can be considered to have concerns categorized as physiological, safety, social, egotistical, and self-fulfillment. Thus, the analyst needs to determine what combination of these factors can be "thrown in the pot" to make the individual left better off than he is under the present system solution. This way, with an individual player there will now be goal congruence. People generally consider money as the primary factor here. But much more imaginative and less costly ways are used by organizations. Gold watches, awards, pats-on-the-back, a call for patriotism, etc., have all been used.

Cost to Overall System

In level 7, we need to establish what the cost is to the overall system of sharing the system benefits. If there is no real cost, i.e., award, a pat-on-the-back, calling the person a vice-president, etc., then the analyst can push for the recommended alternative.

However, as is much more likely, there are some costs to the proposed compromise then the analyst and decision-maker have to assess the overall consequences to the system.

Total Systems Value

In level 8, if after negotiating the compromise the result is a lower total systems value than the existing solution, the proposed compromise should be dropped and the existing solution put forth as the "new best" solution. However, if negotiation results in a total systems value better than the existing system solution, but less than the best total system value possible under the recommended solution, it should be accepted.

In level 9, the decision outcome could therefore be the best alternative, second best alternative, or existing system solution.

DECISION FRAMEWORK EXAMPLE

Because the discussion of the decision framework (Fig. 13-4) may seem a little abstract, let's work through an example to see how it might be applied.

Assume the analyst has been asked by the president of the XYZ Company to do a system study on what the best sales target should be for the company next year.

The analyst, in talking with the CDM (president) and other key decision-makers within the company (i.e., VP sales, VP finance, VP production, etc.) determines that maximizing profit is the organization's major objective. So the analyst in studying the various alternative sales targets concluded that the one with the greatest profit (i.e., measured by return on investment) would be best.

The analyst in studying the situation and the environmental conditions has concluded the following:

Alternatives	Sales Target	Company Profit
Existing system	100,000 units	12%
Proposed solution	90,000 units	17%

Thus, the proposed alternative is better than the existing system because it has a projected profit of 17% as compared to the existing systems profit of 12%. With the proposed solution, the sales target is 10,000 less units than present. However, is it possible to sell less units and make a better profit? (Be sure to consider things like warehouse size, territory, etc.) In any case, assume the figures stated are reasonable.

When the analyst presents his findings to the executive committee (which is chaired by the president and composed of all the V.P.'s such as finance, sales, production, etc.), the president is thrilled with the results. He likes the projected improvement in profit. Most of the committee members like the proposal also, but the VP of sales is giving a lot of flack to the analyst's suggestion.

The analyst listens very carefully to the objections because he knows many people of this committee have much more experience than he does. So maybe he has overlooked something or came to an erroneous conclusion about the profit figures, etc. But the points the sales V.P. makes can be easily refuted. Therefore, what is best for the overall organization is quite clearly the proposed alternative. But there is still grumbling. So what is the problem?

The analyst assumed everyone was an organizational player. But what about the sales V.P.? In previous conversations with other staff members, it was quite clear that the sales V.P. was a concerned individual player.

So the analyst decides to look at the proposed solution through the eyes of the sales V.P. and he sees the following:

Alternatives	Sales Target	Commission	Net Salary	Company Profit
Existing system	100,000	$1/unit	$100,000	12%
Proposed solution	90,000	$1/unit	$ 90,000	17%

So from the sales V.P.'s perspective, he presently is making $100,000 per year. Under the proposed solution, he would be making $90,000 per year. So he is losing $10,000/year in salary. Should he be smiling? If he were an organizational player, he should because his only concern is to help make the company more profitable. In many situations, people respond strictly to the "needs of the service" and disregard the personal consequences.

But in this case, the sales V.P. is an individual player and he is chagrined. But why doesn't he just come out and say that much of the increased profit for the company is coming out of his hide? Possibly the social and psychological pressures of the company may not allow this kind of consideration.

The analyst now begins to understand the confusing rhetoric of the sales V.P. The analyst now sees that this is a situation of goal divergence. What is best for the company overall is not best for the sales V.P. Or it could be that the sales V.P. is willing to go along with the proposal, but the salespeople who work for him are mad because each of them is going to lose $2,000 per year in commissions with the proposed solution. In any case, there is goal divergence.

Now the analyst has to determine the basic power relationships among the executive committee. That is, does the president decide by himself, or is it a case where there is a lot of discussion which the president evaluates before deciding, or is there a vote among all members indicating some kind of consensus, or what? If it is a case where the president decides or there is enough votes in the committee to carry the proposed solution, then per se the CDM has the power to do what he wishes. On the other hand, if the sales V.P. can veto the proposed solution himself, or can convince other committee members using convoluted reasoning (i.e., not saying it is bad for himself directly) not to go along with the new alternative, then the individual has decision power and this must be dealt with.

If the CDM has the power, the analyst needs to consider his management style. Is he one to get goal congruence out of a goal divergence situation by penalty or by reward? The CDM could suggest a third alternative which is to keep everything the same as the proposed alternative but reduce the commission rate from the standard $1 per unit to 80¢ per unit and in this way increase company projected profits from 17% to 19%. The message to the sales V.P. is either go along with the analyst's proposed solution and make $10,000 less a year (i.e., $90,000) or the CDM will push through another solution which is better overall for the company and you will make $28,000 less (i.e., 90,000 units @ $.80 = $72,000). Other possibilities organizations use in the penalize situation are to fire the person, give him a less prestigious job, etc. This all sounds fairly extreme and we are not advocating it per se, but it does take place all the time in most types of organizations (i.e., business, church, university, military, etc.).

A more pleasant approach to goal divergence is when the CDM, with power, decides to use the reward system. This, as it turns out, is the same as when the individual has the power. To get any solution through the committee, this individual must be rewarded (i.e., "bought off") in such a way to result in goal congruence.

What type of rewards should be "thrown in the pot"? Which values to the individual are most important? Assume for now that it is decided to share some of the resulting benefits that would be derived from going along with the analyst's proposed solution.

Alternatives	Sales Target	Commission	Net Salary	Company Profit
Existing System	100,000	$1/unit	$100,000	12%
Proposed Solution	90,000	$1/unit	$ 90,000	17%
Modified Solution	90,000	$1.12/unit	$100,500	14%

Under the modified solution, the increasing of the commission rate to $1.12 per unit but keeping the sales target the same, will result in a net salary to the sales V.P. of $100,500. This is better than he presently is making. What do you think you will hear from the sales V.P. now? "Great plan—always thought we should be selling less," etc.

But there is a cost to the system. This increased money to the sales V.P. had to come from some place. In this case it came out of company profits and the projected profits decreased from 17% to 14%.

But note, the 14% profit under the modified system is still better than the profit with the existing system which is 12%.

So the CDM and the executive committee are now ready to make a decision by selecting a solution for implementation. The three alternatives available are the existing system, the proposed solution, and the modified solution.

The best solution is quite clearly the proposed solution put forth by the analyst, since it would result in an overall profit to the company of 17% which meets the maximum profit objective.

But apparent best isn't necessarily always best. The problem is not what is theoretically possible if everyone were organizational players, but rather what is the best you can, in fact, get implemented. With this perspective, the second best alternative (i.e., the modified solution) is in fact best.

It is better than the proposed alternative because it can be implemented. That is, projections aren't important per se, it is the reality which is crucial. The proposed alternative is better than the existing system, since the company's profit should increase to 14% from the present 12%. Therefore, the CDM should compromise and accept the second best solution.

Let's relook at the situation and assume the negotiations with the sales V.P. under the reward option results in the following situation:

Alternatives	Sales Target	Commission	Net Salary	Company Profit
Existing system	100,000	$1/unit	$100,000	12%
Proposed solution	90,000	$1/unit	$ 90,000	17%
Modified solution	90,000	$1.15/unit	$103,500	11%

In this case, the sales V.P. is demanding $1.15/unit to go along with any change. While it is clearly a good deal for him by resulting in a salary of $103,500 vs. $100,000, notice what happens to the overall company profit. It would decrease from 17% under the proposed solution to 11% under the modified solution. This 11% is less than the 12% profit with the existing system solution. In this case, it appears best to stay with the existing system solution.

However, another possibility is to go back to the sales V.P. and explain that his monetary requirements can't be met because it reduces the effectiveness of the proposed solution too greatly. Then a negotiation phase could take place centered around nonmonetary values. Possibilities could include increasing his vacation from two weeks per year to three weeks, or he could move to a large office with an ocean view, or he could change his title from sales V.P. to the more prestigious (in his eyes) marketing V.P. Or maybe the CDM should go talk to the sales V.P. and tell him that the company needs this profit boost badly and he would appreciate if the sales V.P. would go along with it, even though we all recognize the personal sacrifice he is having to make, etc. In other words, the negotiation phase can be conducted over the full spectrum of Maslow's hierarchy of needs.

Finally, it may be that the CDM, on principle, will not tolerate having to "buy off" individual players who have key power positions. This is his prerogative and what decision-making is all about. The analyst then needs to show the CDM the tradeoffs involved in this regard. In our first example, because the individual has veto power, the company will have to stay with a profit of 12%; whereas, under the compromise position, the company could have gotten 14%. The CDM decides what is best from this viewpoint.

Class Exercise 13-1

WHAT KIND OF PLAYER ARE YOU?

Central to the discussion of how people really make decisions is the concept of the individual player versus organizational player.

1. What kind of player do you feel you are? Give examples from how you carry out your "roles" at home, school, work, with friends, etc.

2. John F. Kennedy, as President, said, "Ask not what your country can do for you, ask what you can do for your country." What kind of player was he asking us to be?

3. What kind of player do the following sayings apply to? "What's in it for me?" "Watch out for No. I." "I want to do what is best for me, which will help the organization."

4. How do people acquire these different orientations? How easy do you think it is to change orientations?

5. In a work situation, is it usually better to have people who are organizational players versus those who are individual players? Give examples where each type of player might be good for an organization and where they might be detrimental.

Class Exercise 13-2

COMPUTER INFORMATION SYSTEM BROCHURE

The following advertisement was designed to make the decision-makers of various organizations want to employ Cal Poly computer information systems students in a work/study program.

Discuss the overall strategy used in the ad in relationship to marketing system solutions. What is the purpose of each section (i.e., *It, You,* and *Why Cal Poly)*? What needs are being appealed to throughout the brochure? Discuss its relationship to the factors affecting the decision process (i.e., organizational player, goal congruence, etc.).

WHAT'S IN IT FOR YOU?

It   is the *work/study program* for students majoring in computer information systems at California State Polytechnic University, Pomona. This program arranges for juniors and seniors to work for a business 15 to 40 hours per week for at least one academic quarter (three months) while they are continuing their formal studies. In addition to being paid (commensurate with their responsibilities) the students also receive college credit for the work.

You   refers to you, *businesses using computer processing* in the southern California area which can also benefit from participation in this work/study program.

How? You have the opportunity to:

1. Have top quality, highly motivated young men and women join your staff

2. Have first crack at the top level graduates in the field

3. Enhance your company's image with the community

4. Get top quality performance for less salary

5. Get fresh ideas from young people still in college

Why Cal Poly?

If you are interested in the idea of a work/study program, you will be interested in knowing why involvement in the Cal Poly program would especially benefit you. Most important among the advantages are:

1. Computer information systems courses at Cal Poly are concerned with solving real-world business problems, not just theory.

2. Participating Cal Poly students will be near graduation, with most of their required courses in data processing completed.
3. Cal Poly students concentrate in both systems and programming applications.
4. In addition to computer courses, Cal Poly students get a broad business education which includes several courses in accounting, marketing, finance, management, and economics.

Class Exercise 13-3

AFFECTING DECISION CONSIDERATIONS

1. Why did the following employees go out on strike when they did? Baseball players struck one week before the start of the major league season. Airline clerks struck at the start of the Christmas season. Lettuce growers face bracero's walkout at the peak of harvest time. As a tactic, how does this differ from unions which suggest a sickout or "blue-flu"? What "type of management" would state that employees who stay out will be fired? That is, what is the management style?

2. In evaluating how decisions are made, why are there so-called pork barrel projects? Why are there so many military bases located in the home state of previous chairmen of the Senate's Armed Service Committee?

3. With the advent of microcomputers, the users of computer services in many large companies are demanding their own computer power. Why are some DP managers very fearful of this trend, while other DP managers welcome it? Conversely, DP managers sometimes want to go to decentralized computer usage but the potential users want no part of this setup. Assume in either case, that technical and economic considerations support the particular thrust. What else must be considered?

Chapter Fourteen

Implementing
the System Solution

In the previous chapter, we saw how people approach decisions and why additional factors must be considered beyond the "rational" analysis, to get an alternative chosen. However, just because a solution is chosen, this does not imply it will be successfully implemented. In this chapter we will discuss who makes up the acceptance groups. Further, there will be definitions of the major factors affecting solution acceptance. How to use these factors to increase the chances for successful implementation will be discussed. Lastly, examples of how to implement change will be given.

ACCEPTANCE GROUPS

There are four main groups which greatly influence the full implementation of the recommended solution. These groups may be identified as designers/analysts, preliminary/decision-makers, CDMs, and users/workers. Together they will be known as the acceptance group. Each group enters the implementation process at different stages of the acceptance process.

The designers/analysts are generally the first group to formulate the alternatives and make an initial evaluation. Through them, the alternatives are given a relative ranking. It is very important to appreciate that the analyst must be as unbiased as possible in the predecision phase of the system study. It is only after the alternatives have been evaluated that the analyst has a choice of acting as a salesman or a reporter.

If the designer/analyst chooses to act as a reporter, then he is not a factor in the acceptance process. However, if he feels it is in the best interest of the organization to act as a salesman of the "best" solution, then his perspectives and relative influence become a factor in the eventual success of the systems change.

Preliminary decision-makers affect the implementation process in the sense that they have implicitly the power of veto. That is, if the legal advisor to the company says the alternative, as formulated, cannot be accepted, then most often the proposed solution is dropped or greatly modified to make it legal. A similar process is undergone with the VP marketing, the controller, etc. These people can say no, and the alternative is dropped. But they can't say yes in the sense of implementing the solution. Rather, their concurrence means it looks OK to them. The final decision on implementation is still with the CDM.

The CDM, by definition, and in fact, is the person or committee who can decide which alternative is to be implemented. It could be the "best" alternative, "second best" alternative, or to stay with the existing system. Considerations involving these possibilities were discussed in chapter 13.

But it is very important to appreciate that just because the CDM says "implement," this does not imply that the recommended solution will be fully incorporated into the system. There are two additional points to be considered. First, there is a tendency of human beings to want to put things off to tomorrow instead of doing things today. So even if the people who must implement the solution feel this alternative is in their best interest, there still must be plans and procedures to see that things are carried out in a timely fashion. The second point is, what if the users/workers (implementors) feel the solution that is to be implemented, has worse consequences for them than the existing solution? Depending on whether they are organizational players, how much power they have, etc., a certain degree of resistance will be formed.

In relatively simple systems, one person can be the analyst, CDM, and user of the proposed system solutions. In more complex systems, each of the acceptance groups is represented by large, sophisticated groups (i.e., the development of a new fighter aircraft for the military). For our purposes, we need to understand the varying needs and requirements of the acceptance groups at the different stages of the problem-solving process.

MAJOR FACTORS AFFECTING SOLUTION ACCEPTANCE

While there are a multitude of factors which affect the acceptance of a proposed system solution, we will group them into four major categories: pressure, relative advantage, goal congruence, and amount of behavioral change.

Pressure

The more the present ways of functioning are in trouble, the more likely that the acceptance group will accept new techniques. The greater the difference between expected and actual performance, the more pressure there is to do something. The pressure increases when the deviation is occurring today as opposed to a future time period. The more the present system is not viable, the more the pressure.

Thus, operational difficulties generally create more pressure than do strategic problems (though it should probably be the other way around). When there is a gasoline shortage, such that you can't get to work, you have considerable pressure to do something now; whereas predicted gasoline shortages of five years from now generally provoke just a smile and a nod.

Much of the real innovation in the military has come about during the crisis of war. The pressure of survival has forced new solutions to be tried. For example, during World War II, radar was introduced in Britain as a last-ditch measure to help reduce the effectiveness of aerial raids by Nazi aircraft; the Germans introduced the jet fighter airplane late in the war in an attempt to regain air superiority; the United States built and used the atomic bomb to hasten a victorious conclusion to the war; the Japanese introduced the kamikaze tactics as a last try for victory.

Crises force change in a search for solutions. If things are going well, there is generally no pressure to change. For example, if a business is very profitable and has been for the last ten years, why should it switch its methods of operation to try new solutions? Compare this situation with that of a compnay which has lost money continuously for the last five years. Because its very survival is at stake, it is more likely to try something new.

Relative Advantage

The greater the total value of the proposed solution to the value of the existing solution, the more readily the new solution will be accepted. A solution which is 90 percent more effective than present methods will be more likely adopted than another proposal which is only 10 percent more effective. This assumes of course that the effectiveness numbers are valid. However, the degree of relative advantage is always discounted by the acceptance group to a degree based on how believable the alleged advantage really is.

When a computer salesman comes to an organization and says he can reduce the cost of computing the company payroll by 50 percent, the company CDM would very likely switch systems if it were in fact true. The more proof of the relative advantage given, the more believable it is, the more likelihood of change.

The comparison of the proposed alternative and the other solutions is given in the evaluation matrix (chapter 12). Recall that the alternatives, including the existing system solution, were compared on the basis of total value and on discounted value. In addition to the relative value of the preferred solution over the existing solution, the more the total value and discounted value both show the proposed solution is better than the existing solution, the more likely the implementation.

Goal Congruence

This expresses the degree to which the proposed solution furthers the goals of the affected groups such as the analyst/designers, preliminary decision-makers, CDM, and users/workers. The reasoning and considerations behind this topic were discussed in depth in chapter 13. We will only recap the main points here.

For each group, when acting as individual players as opposed to organizational players, to go along freely with a proposed alternative, they must feel they will be better off than with the existing solution. If they are not, they will fight the acceptance process.

If the affected group has the power to veto the proposed plans, then the proposed solution most likely will be modified to bring about greater goal congruence. While compromise and accommodation increases goal congruence, it also lowers the overall relative advantage of the modified solution over the existing system solution.

Behavioral Change

The last major factor is the amount and type of behavioral change that is required of the affected parties. This will be discussed in terms of the group's inherent attitude toward change, and the amount and type of change required.

Innovativeness of Acceptance Group: Everett Rogers has shown in his classic book, *Diffusion of Innovation* (1962), that people are very different in their reaction to new innovations. He has set up a scheme in

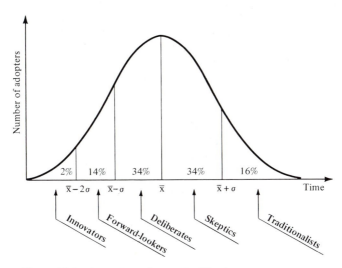

Figure 14-1 Acceptance Group classification

which people are classified as innovators, forward-lookers, deliberates, skeptics, or traditionalists depending on the relative length of time before they fully accept new concepts (Fig. 14-1). This scheme stresses the idea that people have *markedly* different desires with respect to trying new concepts, products, system solutions, etc.

Innovators are those people who first accept an idea. They have almost an obsession to try new things. These venturesome people are daring and possibly rash.

Forward-lookers constitute the early adopters who are generally among the first 16 percent of the people who adopt an innovation. These people are generally not obsessed with trying new things, but are constantly looking and evaluating better ways to do things. They are generally well respected for their judgment on new ideas.

Deliberates are those people who constitute the early majority of new system adopters. Their general motto is "Be not the last to lay the old aside, nor the first by which the new is tried." They tend to watch new developments in a positive frame of mind.

Skeptics are those people who constitute the late majority of adopters who approach innovations with a cautious air. They generally will not adopt an innovation until public opinion (or other companies, executives) definitely favor it.

Traditionalists are those people who are the last to accept new ideas. They are quite suspicious of change, and tend to favor the status quo very strongly. Their attention seems to be fixed on the past rather than the new and modern.

Thus, we see that people vary to a very large extent in their attitude toward change. However, it is not true that if a person is classified as an innovator that he is an innovator in all things. For example, a businessman might be quite an innovator at work in his attitude toward new ideas and change, and at the same time hold very traditional values toward his family and marriage. A woman could be avidly fashion-conscious and wear all the newest styles (innovator), but at the same time at work drag her feet in regard to any change in procedures.

For the purpose of the system study, only the attitude that generally prevails in the acceptance group toward change in the system understudy is important. The people at work who are the innovators and forward-lookers will much more readily accept an innovation, than people who are generally highly skeptical toward change and will not move until excessive proof is shown.

Amount and Type of Change: In general, the less behavioral change required of the acceptance group, the more readily the new system will be accepted. When the proposed solution is just a minor modification to the existing system, there should be little reluctance to accept it. An example would be an improved system solution which kept the same basic system but removed some of its disadvantages. The switch over from hand-crank calculators to electronic calculators for faster, more efficient bookkeeping operations should be quite readily accepted.

A major alteration of the present system requiring significant behavioral change by the acceptance group will be harder to get implemented. The automation of many machine tooling jobs allows operators to function more effectively than before. However, accompanying this increased system effectiveness is great behavioral change required on the part of the operators. We wouldn't be surprised if they resisted the change, even though their wages were increased. These people will be required to learn new skills to operate the machines and may also suffer a reduction in status.

The greatest change is required when the proposed solution requires a completely new and potentially revolutionary way of doing things. An example of this has been the introduction of computers to most types of organizations. The computer normally requires tremendous behavioral change within the organization. New organizational structures, power/knowledge relationships, decision requirements, etc., are brought about. To the computer experts the innovation is fantastic, but to all the others who don't understand computers or who are being shifted aside, it can seem confusing or even threatening.

Thus, in general, the more behavioral change required, the less likely people will go along with it, with traditionalists being the least likely. In addition to the amount of change, there is the type of change required. The more the change attacks or removes the fundamental values of a person or a group, the more resistance to the change. That is, considering Maslow's hierarchy discussed in chapter 13, if the proposed plan attacked the survival needs of a group, you should not be surprised that they would use even violent means to keep proposed solutions from being implemented.

An example of this last situation would be a proposed alternative to replace the machinists in the plant with an automated computerized milling machine. Since the livelihood of the machinists is being threatened, and their job opportunities at other plants are limited, you could expect a rather violent strike calling for complete rejection of the computerized scheme.

Further, the more people are accustomed to doing something in a certain way, the harder it will be to change them. The recent campaigns against smoking are a good example. If a person has been smoking three packages of cigarettes per day for twenty years, it is going to be very hard for this individual to change. Even though the person would admit that smoking is hazardous to his health, he may have formed a very fundamental habit which will be hard to break. This concept can be summarized by a modification of an old saying, "You can't teach an old dog new tricks." While age is somewhat important, the more critical factor is how long a person has been doing something a certain way.

Summary

To summarize the major factors affecting the acceptance of a system solution:

The more *pressure* on the acceptance group to make a decision, the more likely the group will be willing to try something new.

The greater the *relative advantage* the recommended solution has over the present system solution, the more likely it will be accepted.

The greater the *goal congruence* among the acceptance groups, the greater the likelihood of the new solution acceptance.

The better the match in the *behavioral change* required to that which is desired, the more likely the new solution will be implemented.

PLANNING SYSTEM CHANGE

In the previous section, the major factors which affect the acceptance of a solution were discussed. We now consider how to use these same factors to increase the likelihood of solution acceptance by planning the change to the system. This can be brought about by creating pressure, proving the relative advantage, increasing the goal congruence, and decreasing the behavioral change required.

Creating Pressure

It was stated that the more pressure that is on a group to change, the more likely it will be willing to try something new. If a group is facing a crisis situation where the present methods are inadequate, there is a good chance of introducing new methods. Whether crisis is inherent in the situation or is intentionally initiated, it will tend to have the same effect—change.

Examples of crisis inherent in the situation would be when a company's competitor has launched a breakthrough product (Radio Shack's microprocessor in the computer industry) or where the underlying business is failing (passenger train business—Penn Central merger).

Crisis can be deliberately initiated. How crises can be intentionally initiated depends on how much control the changers have over the potential acceptance group. At one extreme would be the Pentagon deciding to implement a new airplane weapon system into the military services, where they have close to full control over the acceptance group through authority, budgeting, and tradition. At the other extreme of control would be young anti-war activists with no internal authority attempting to get Dow Chemical Co. to quit selling napalm to the military during the early 1970s.

When there is close to full control, the CDM can call his key people together and explain to them the importance of the project and his desire for a smooth implementation. This explicit CDM support for a program will have the effect of putting pressure on key people. If this initial support is combined with a good system for monitoring actual program change, it will help to ensure an effective program implementation. Further, careful attention to project control by developing an overall implementation plan, setting deadlines, establishing milestones, and designating specific people to be responsible for particular tasks is generally very effective, since it applies pressure where and when it is needed.

When the change agent has no control, how does he put on pressure for change? What is one of the major tactics of groups like Nader's Raiders, Chavez's United Farm Workers, etc.? It is to put the group or organization they want to accept change (their system solution) under tremendous pressure. This pressure develops when they make the present way of doing business by the organization no longer workable. Crisis can be brought about through conventional "channels" like convincing stockholders in an annual corporate meeting to vote against certain actions or it can be brought about by anarchy for example by burning down the Dow Chemical plant for its continued indirect involvement in the Vietnam War. At either extreme, the purpose is to get change.

Increasing Relative Advantage

The better the new system solution accomplishes the system objectives as compared to the present methods, the more likely it will be implemented. Aside from the obvious idea of trying to think up an even better solution to get greater relative advantage, what can be done to increase the probability of acceptance?

One of the problems hindering new solution acceptance is the real doubt as to whether in fact the new solution can do everything claimed for it. The claims of salesmen, consultants, and analysts are notoriously overoptimistic and thus the acceptance group may discount the claims so heavily that the new system may seem to have no real advantage. Therefore, a change agent can increase system acceptance by giving proof of what the system can do.

Depending on the circumstances, showing the relative advantage of a system can either be very simple or almost impossible to accomplish. A simple proof is when a life insurance salesman works out a specification sheet which shows how his company can give you the same exact insurance coverage as you presently have at a 30 percent discount in price. A much harder proof involves the move in Congress to establish an Office of Technological Assessment in the late 1970s. The purpose of this office is to limit technological advancement to only those developments which will have a net social effect which is positive. While most would agree with the value of this idea, trying to prove the actual worth of this agency before the fact is essentially impossible.

In any case, proof should be given to verify the relative advantage of a system solution. In simple cases, proof can be given along with the initial presentation. For those situations where proof is essentially impossible or too costly to obtain before the fact, a pilot implementation can be initiated. This pilot study can be used to check out whether or not the systems results can really be attained or not. It also provides a chance to make modifications in the system under actual conditions, so to insure or even improve system effectiveness.

The other tack is to show that the actual value of the existing system is much lower than people commonly think or to show it will be undesirable over the future planning horizon. Much of this can be accomplished by a good briefing or short written document which delineates the reasoning behind the analyst's thinking and what other sources of information agree with these conclusions.

Increasing Goal Congruence

The previous discussion centered on the relative advantage of the proposed system solution from the organization's viewpoint. But what about the individual user himself? Is it reasonable to assume that a group of workers will readily accept automation because it is good for the company, even though they lose their jobs? Will a person be willing, for the overall good of the organization, to take another position with the firm which pays the same where it is clear to all that he has a lower status? A very important factor in how readily change is accepted is what is the net effect on the individual user? This effect must consider more than simply economics. It must also include psychological and social factors.

The more a person gains individually from a new system solution, the more readily he will accept it. Conversely, the more an individual loses personally from a new system solution, the more he will tend to resist system introduction.

The ideal situation in implementing a new solution would be not only for the organization as a whole to benefit, but also each individual connected with the system change to be left better off. Usually this is almost impossible to bring about. The change agent needs to assess which groups are going to lose and what can be done about that to increase the goal congruence of the groups.

To ensure that a system solution will be implemented, a less effective solution from an organizational standpoint may be accepted. The use of compromise to appease the various groups (individuals) by giving them a net positive return (or at least less negative) by accepting the new solution is well known. The example of introducing the much more productive diesel engines to the railroads illustrates this point. From an organizational point of view, the diesel engines had a very great advantage over the steam engines. Therefore, the railroad desired to rapidly replace all steam engines with diesel engines. From the worker's viewpoint, the now redundant fireman has everything to lose from this innovation and nothing to gain. As a compromise to the unions to get the diesels in, the railroads accepted featherbedding (i.e., firemen on diesel engines). Because the firemen are not needed on diesels, this results in a less effective system solution from the organization's viewpoint. However, by accepting a less effective solution, the firemen are left in a positive position and the system can be implemented.

Decreasing Behavioral Change

The factor that the change agent seems to have the least power to influence directly is the innovativeness of the acceptance group. The particular innovativeness that a person has along certain dimensions is a function of his total life experience. It is not likely to be changed rapidly. The innovativeness of a group or organization is a function among other things of its mission, its tradition, and the people it is composed of. While many organizational development (OD) people feel they can change people or group attitudes toward change quickly (an extreme claim is a one-week seminar), most of the evidence suggests it cannot be done so quickly or permanently, if at all (Richards and Greenlaw, 1972).

Assuming the above is true, then the innovativeness of the acceptance group is essentially a given condition. If the group is traditional in their outlook, they will be much slower to accept any new system solution, whereas if the group is forward-looking they will accept change much more readily. All the change agent can do then is to assess what the innovativeness of the acceptance group is and see how this will help or hinder new system introduction. If it will hinder the change-over, the analyst can't do anything to affect the group response directly.

There are three things the change agent can do to indirectly change group innovativeness: (1) exchange group members, (2) establish a new group, or (3) increase group acceptance. If some members (traditionalist) of the acceptance group are transferred out of the group and replaced by new members (innovators), the net balance of the group may be toward more innovativeness. In extreme cases, the group can be disbanded and a new group formed or the system solution can just be given to some new group to be responsible for from now on. In either case, the group innovativeness has been increased. President Roosevelt used these tactics in the 1930s to get his New Deal program implemented. If he had tried to work these new programs through the highly traditional bureaus, the programs would have been suffocated. So Roosevelt just formed new organizations and gave them the missions.

A third way a change agent can increase group acceptance is not to try to get the whole group to accept at once, but to appeal to various group members at different times using different tactics. The first members to appeal to are those who do already or will most readily favor the system solution (innovators and forward-lookers). Possibly through a briefing, and a question-and-answer session these members will be able to see the value of the program and be willing to support its implementation. To move the members considered deliberates, there will need to be proof of the plan's value. If this is not easily attainable, then a pilot implementation plan should be suggested. Through a pilot program, proof can be given of the system's worth. If in fact that system does what was claimed, those members who are considered deliberates will now be inclined to support the program.

At this point the change agent is ready to tackle the members of the group who could have initially killed the program—no matter how good it was! To get the members who are skeptics or traditionalists, and prefer the status quo rather than going along with any new program, the agent has to be able to put group pressure on them. Since there is now proof of what the system can do, the change agent should be able to neutralize any significant argument they can make favoring the status quo. Note, if this cannot be done, maybe the system plan is not that good after all, and the organization should stay with the present system solution.

Once the point is reached where the majority of the group favors the plan, it becomes the accepted thing to favor. To be against the plan now requires withstanding group pressure. Since the arguments of the skeptical members have been effectively countered, they have nothing really to stand on and will eventually switch, or in many cases can just be ignored if they still won't switch.

Amount and Rate of Change: The more the system solution proposed is compatible with the existing values and past experiences of potential adopters, the more likely it will be accepted. The more behavioral change involved, the harder the system solution will be to accept.

The amount of change *perceived* by the acceptance group is what is important here, not just the actual change. When there is considerable uncertainty about how a program of importance is going to affect people, the perceived change can be seen as much larger than the actual change. A typical case is when people are anxious about what will happen to them and others because of a proposed new program. A rumor mill starts and with each retelling the story acquires distorted and exaggerated claims as to how the new system will affect them. Since people react to the amount of perceived change, the new program will be much harder to get accepted because of exaggerated and distorted views of the amount of change required.

An effective way of bringing the perceived change more in line with the actual change is to explain the implications of the new program. Explanation should include what the purpose of the program is and how it will affect different acceptance groups. This explanation should be given by a knowledgeable person through meetings, rap sessions, and/or written documentation.

In those cases where the total change required of people can't be reduced, it can be made more palatable by introducing the change in smaller, more readily accepted steps. For example, a pilot implementation program gives people more time to accommodate to the required change. An announcement one morning to all company members that the total accounting system is being switched over to a fully computerized system will surely result in chaos and much pain. Much more acceptable would be the announcement that a trial program is being set up to see the effect of computerizing the accounts receivables of the company accounting system. If that proves successful, implementation will proceed to inventory accounting, etc. While the total changeover is eventually the same (total computerization of the company accounting system), the smaller step approach will have a much better chance for implementation because people will have more time to adapt and thus perceive the change to be less radical.

Ease to Use and Understand: The easier a program solution is to use or put into effect by an acceptance group, the quicker it will be accepted. Generally speaking, the more a group understands how the process works, the easier it will be to accept change.

One of the major ways to get groups to increase their understanding of a significant program change, is to have some kind of educational seminar. The purpose of the seminar is to increase the skill level of the acceptance group to a point where they can both understand the program and learn how to use it effectively. Documentation of a program provides another means for people to understand how the innovation works and become proficient in its use. While documentation is generally not as effective as educational programs, it can be far less costly to provide.

For programs of major importance and/or significant complexity, an on-the-job expert needs to be provided to the acceptance group. The full significance of the program and how to make it work effectively cannot be learned totally through educational seminars and program documentation. Someone is needed to provide expertise on how to get the new system through the rough spots. As the acceptance group gains sufficient know-how to keep the system running effectively, the loaned expert can be moved to other groups without seriously jeopardizing the successful system implementation.

HOW TO IMPLEMENT CHANGE—EXAMPLE

The factors important for implementing change have been discussed in some detail. Which factors to stress and how specifically to address them varies with the situation and depends on what the change agent is trying to get implemented. Several examples will be discussed to try and give a better idea of what is involved and how to go about it.

Example 1

A manufacturing company had ordered a full study to determine whether it is better to computerize their complete accounting system or keep the manual accounting system as it is now. The analyst's recommendation to the president of the firm (CDM) was to fully computerize the accounting system. The president agrees and now wants an implementation plan drawn up. What would you do?

Change Agent's Recommendations

1. CDM support for plan
2. Small scale implementation program
3. Start with high probability of success areas first
4. Assign program manager
5. Establish goals, deadlines, and measures of success
6. Educational seminars

Change Agent's Rationale: The company has been quite successful during the past 10 years and is under no pressure to make any changeover in procedures. However, this changeover will make the organization even more efficient than it is now. Because there is no inherent crisis pushing for change, pressure needs to be generated. Therefore, the change agent recommends the CDM call his key people (staff, function heads, and workers) together and tells them that he feels this project is important and explains the necessity for a smooth, successful implementation.

To both verify the benefits of the switchover and to lessen the rate of behavioral change required by various groups, the analyst insists on a small-scale implementation program. While there are many places that this program could be started, he feels the greatest chance for initial success is with the cash management section. This is a relatively straightforward application and Mr. Smith, who is in charge of this section, is very much in favor of the switchover. The president, however, had recommended starting with accounts receivable.

In talking with Mr. Jones, who heads that section, the change agent noticed a great reluctance on the part of Mr. Jones to change (a classic status-quo type), and he would most likely do his best to torpedo the program. Therefore, it has been suggested to approach him last after much program success has been built up.

It is definitely recommended that a program manager be assigned to have full responsibility for getting this system implemented. Goals, deadlines, and measures of success must be established. Since about half of the accounting section knows nothing about computers, there is need for a low-level educational seminar. The purpose of the seminar will be (1) to try to remove their fear of the computer, (2) to instill a positive attitude toward the switch by showing how it will help them individually, and (3) to show them how to use the specific computer system in question.

This type of plan will help ensure the successful implementation of a new program.

Example 2

A traveling sales representative of the ABC Company made a request through the sales manager to the president of the firm, that the company cars be equipped with air conditioning. The president, in turn, had his analyst take a look at the cost-effectiveness of this request. The analyst's recommendation was that on a strictly economic basis the air conditioners could not be justified. However, if psychological factors were also considered (like morale, pleasant working environment, etc.), then the air conditioners would be of value both to the salesmen and the company. The president agreed and asked for an implementation plan. How would you implement this system change?

Change Agent's Recommendations

1. Authorize sales manager to implement
2. Set limits

Change Agent's Rationale: Since the system change is very straightforward, of minimal cost, and desired by the users, the change agent recommended letting the sales manager and his salesmen work out the details of how they want to implement the plan. In this way they can decide what is best for themselves in terms of which cars to air-condition first, who should get these cars in relationship to sales schedules, etc.

The agent also recommends putting an upper limit on the amount that can be spend ($5,000) and requiring that all company cars be equipped within a certain time period (12 months).

Overall, the cost of being wrong in this situation is so small, it is not worth much elaborate planning or monitoring.

Example 3

Assume it was 1979 and the "oil crisis" hit and you were an analyst for a major oil company. Further, you were asked how you would convince the American public to be happy paying $1.50 per gallon of gas when they had been paying $.70 per gallon. How would you implement this solution?

As you start to look at the situation, you see you are in an individual player situation. The oil companies and their customers are different entities. So you then look for goal congruence. The doubling of the price for a gallon of gasoline is good for the oil company, but certainly isn't good for the consumer. So there is goal divergence.

Since you are dealing with a basic situation of goal divergence with an individual player, your hope for ready acceptance of the oil company proposal to raise the price of gasoline to $1.50 per gallon is sure to be resisted. Further, the consumers have the power through their votes in Congress to bring heavy pressure on the oil companies.

So you decide on another approach. You meet with the President of the United States, the Department of Energy, and certain powerful senators and educate them to the realities of the gasoline situation caused by the OPEC cartel.

From this group, comes the announcement that the greed of OPEC is causing the Free World serious problems, and cheap energy is now a thing of the past. Gasoline rationing may be necessary and prices of $2.00 or $3.00 per gallon could become common. This announcement forces the U.S. consumer to vent frustration against those "foreigners" (OPEC), who the consumers know we don't control. This, in turn, lessens the pressure on the oil company presidents, the senators, etc. Secondly, the trial balloon approach of seeing the U.S. consumers' reactions to a very drastic proposal ($3.00 per gallon and rationing) helps the change agent to see which way to proceed.

But to make sure the consumers are convinced of the crisis, the oil companies deliberately divert oil into other areas and cause a shortage, which requires people to wait three and four hours for gasoline. After about a month of this, consumers who have been put under tremendous pressure to get gasoline to continue to go to work, are very willing to pay more money to just get gas. Then when the oil companies announce that the price will be $1.00 per gallon with no rationing, it looks good by comparison (a version of the penalty approach from those in power).

By easing the price up 10¢ per gallon each month, as opposed to an immediate doubling, the rate of change required is lessened to a more readily acceptable rate. But in case this doesn't make the consumer happy to pay $1.50 per gallon, the President of the United States goes on television and tells the American public there is a grave danger to our way of life, he is declaring the moral equivalent of war on the energy situation and it is the highest degree of patriotism for you to decrease your gasoline usage and be willing to pay whatever price that is required by the oil companies.

Patriotism and the moral equivalent of war are ways of making inherently individual players into organizational players. In this way, there is no real concern for goal divergence, since the organizational player experiences goal congruence by asking what he can do for his country.

This fictitious gasoline example was used to show how in more complex system situations, the various implementation factors could be used to induce change in a system. Its relationship to the actual 1979-1980 oil crisis is unknown and unimportant for our learning purposes.

Overall Concept

The overall concept to be derived from the above examples is that the plan for implementation must be tailored to the situation. Sometimes there is need for elaborate plans, program managers, documentation, tight controls, while at other times it is best to give only some general guidelines. The successful analyst learns the difference.

Class Exercise 14-1

DETERMINING YOUR INNOVATIVENESS

1. The innovativeness diagram (Fig. 14-1) classifies people according to the relative rate at which they adopt new ideas, products, etc. Rate yourself on the following items according to whether you were (1) one of the first, (2) before most people, (3) about the same as most people, (4) after most people, or (5) one of the last *to adopt* this particular idea, product, etc. Consider only those items which you have adopted.
   (a) jogging
   (b) health foods
   (c) latest fashions (hair, designer jeans, etc.)
   (d) against nuclear power plants
   (e) bought a water bed
   (f) use seat belts in your car
   (g) feel there is an energy crisis
2. What do you think? Are you venturesome or a laggard? Is that good or bad?
3. Will Durant, a famous historian, has stated in his book, *Lessons of History*, that he feels that 99 percent of the new ideas presented throughout history have not been as good as the systems they replaced. Whether he is correct or not, let's give some support to his views by saying that all new ideas are not necessarily good.
   (a) If this is true, how do you feel now about people who resist change?
   (b) What do "can-do" people implicitly assume?
   (c) Does this assumption make a case for advocacy under conditions other than certainty?

Class Exercise 14-2

IMPLEMENTATION CONSIDERATIONS

In the following examples, identify who the crucial acceptance group is, and discuss the inherent pressure, relative advantage, goal congruence, and behavioral change that is implied. Lastly, develop a plan for each situation for increasing the likelihood of successful implementation.

1. The United States is one of the last countries in the world to accept the metric system. Why is this so? Why has so little progress been made in this regard in the last twenty years?

2. In the late 1970s, supermarkets tried to take advantage of the latest technology by introducing optical scanning devices hooked to computerized cash registers. By utilizing this new concept and the bar codes on supermarket items, the grocery industry felt they could save money and speed up the checkout process. What has happened with this experiment? Look at it from the viewpoint of the supermarket executive, customer, computer salesman, etc. In a test market situation, one grocery company introduced the concept in Florida and it was a disaster. But the market test in a major college town in California was very successful. Why?

3. Because of the uncertainty surrounding future oil supplies and the increasing price of that petroleum which is available, the time seems right for the electric car. If you were working for a car manufacturer who had a viable electric model, how would you introduce this concept to the American public?

Chapter Fifteen

Communicating Study Results

In this text, we have presented an integrated approach for systematically solving complex systems problems. We are now at the point where the analyst needs to communicate his findings to the various decisionmakers, and to get a decision as to which solution will be implemented. The fundamental concepts of communication will be given as they relate to the tasks required of the systems analyst. When an analyst should be a reporter and when he needs to be a salesman will be discussed. How to prepare for a presentation and what strategies to use to accomplish specific objectives will be given. In this regard, a knowledge of the various decision styles people employ is very helpful. Lastly, the general format for communicating the study results will be explained for either a verbal or written presentation.

CONCEPTS OF COMMUNICATION

How to communicate what has been done and to show the results of the systems study, is a very important part of the systematic systems approach. To communicate effectively, the analyst must consider (1) what role he should be playing, (2) the purpose of the communication, and (3) the audience characteristics. From this, the analyst can determine the best method of communication. A summarization of these concepts appears in Table 15-1. Finally, to make all these facets more effective, the analyst needs to have a good understanding of some of the basic laws of communication.

The first question the analyst needs to ask when he is considering communicating the results of the system study is, What is the overall purpose of the presentation? Is he simply reporting to others what he has found out? Or is he giving a progress report of what the present status of the study is? Perhaps the analyst is trying to educate the prospective audience about the underlying subject matter. Lastly, the analyst's major purpose could be to persuade the CDM and others that a particular solution is best. The purpose of the communication influences the way the analyst should proceed.

The next major question that needs to be addressed is, Who is the audience? Potential audiences are the designers/analysts, preliminary decision-makers, CDM, and users/workers. As will be covered in more detail under laws of communication, the analyst needs to communicate in a different way with different audiences. As an example, fellow analysts probably need much detail and many specifics, whereas the CDM wants an overview of the situation with multiple perspectives. Preliminary decision-makers most likely want a brief overview of the study, but are interested in a more specialized perspective (i.e., the impact on production) with detailed information on this dimension. Workers need to know the details of how it affects them.

Thus, beyond what is to be communicated, the background/ knowledge of the audience influences how the study results should be communicated. In addition to considering who the audience is, the analyst needs to determine how many people need to be communicated

Table 15-1 Concepts of communication.

Purpose
    Information
    Persuasion
    Progress
    Education
Audience
    Who is it?
    General background/knowledge
    How many are there?
Major Methods
    Written
    Verbal

with. In general, the more people there are, the less the communication can be tailored to specific individual needs.

The major methods of formal communication are written and verbal. Written communication usually refers to a typewritten document which could be in the form of a memo, summary pages, or a detailed report. Further, these can come in a draft or final copy. Verbal communication revolves around a formal, oral communicating of the results. This can take the form of flip charts, vu-graphs, and/or slides. The larger the audience and the more formal the presentation, the more slides should be used. Flip charts are a very handy and effective means for the analyst to give a relatively small audience a briefing of the study results. Vu-graphs are a compromise between the slides and flip charts. These are very easy to put together and can be viewed by a relatively large audience.

Laws of Communication

To make an effective presentation to a particular audience, for a specific purpose, communicating in a written or verbal form, the analyst needs to understand some basic laws of communication. As shown in Table 15-2, these laws are divided into considerations of perception concepts, framework of the audience, effective communication, and reading the audience.

In general, one only communicates with an audience when that person or persons receives, then understands, and lastly, is influenced by what is said or written. Some of the senses are better attuned to communicating information than others. Studies have shown that seeing something is more effective than just hearing it. Further, seeing

Table 15-2 Laws of communication.

Perception Concepts
    Some senses more effective than others
    More senses used, the better
    Medium can blur message

Framework of Audience
    Start where they are
    Translate concepts into their language
    Move at their pace

Effective Communication
    Introduce topic, discuss topic, then summarize topic
    Present advantages/disadvantages to all solutions

Reading Audience
    Beware of nonverbal clues
    Questions are an indication of interest and understanding

and hearing something is more effective than either sense alone. Where practical, the analyst should try to take advantage of the fact that the more senses that are involved in receiving information, the more likely it will be understood.

The format in which the information is sent can very heavily influence how effectively it is received. Using Marshall McLuhan's terms, the medium is the message. Hanan, in his book *Sales Negotiation Strategies (1977)*, has stated that effective communication is not just saying the right thing in the right way. This is important, but not the paramount aspect. Studies have shown that only 7 percent of face-to-face communication is accomplished by words. Vocal intonation and inflection account for 38 percent. Facial expression and body language account for 55 percent. In other words, the manner in which something is presented is often more important in determining what information is received than the information itself. As was shown in Chapter 11, the evaluation of information is generally a mixture of the information itself and an evaluation of the source.

The analyst needs to tailor his information to the characteristics of the audience. In a verbal briefing where the analyst gives the presentation to a vice-president of the firm, conservative business dress is recommended. Formal written studies should be typed and attractively packaged. In other cases, the analyst needs to talk to a worker on the assembly line at his or her level (i.e., not down) and, therefore, should probably be more casual in dress and manner. The most general comment is, "When in Rome, do as the Romans do, but be slightly more conservative." The overall point is, if the CDM is concerned about your outlandish tie, he is thinking of only that, and not listening to your talk.

Once the information is presented, there is no guarantee that it will be understood. To increase this probability, the analyst must think in terms of the audience's framework. It is not enough for the analyst to understand what she herself is saying, it must be understood by the audience.

To do this, the analyst needs to translate her thoughts into the language of the audience. Important points can be conveyed by stories or examples. For instance, if the audience is familiar with cars, then make your points in terms of cars.

The analyst should always start where the audience is in terms of background about the subject matter. Then it is a case of progressively taking them to the final point you want to reach. The subject matter and their expertise should be used as a guide to the pace at which the analyst should proceed. "Bite size" chunks are best. If the analyst moves through the discussion too quickly or at too slow a pace, the audience has a low probability of arriving at the same end point or conclusion that the analyst desires.

For any person to have a good understanding of a situation, he must be shown multiple sides to the question. In our case, this implies presenting the advantages and disadvantages to all the relevant solutions. At a minimum this would involve comparing the recommended alternative with the existing system solution. Further, as a means of effective communication, you need to tell the audience what you are going to tell them, tell them, then tell them what you told them. The first part is generally a short orientation to what will follow. The main body of the talk or written report is to provide detail about what the problem is, what can be done about it, and the way to go. The conclusion is a short summary statement and a call for action implementing a particular solution.

The analyst needs to be able to sense how effective his communication is. Nierenberg and Calero in their book, *How to Read a Person Like a Book*, suggest various types of nonverbal communication that can give the analyst insight into the significance of gestures or body language. Very importantly, these authors feel that clusters of gestures (i.e., the whole body not the individual parts) must be used to get reliable indications of a person's attitude. From these clusters, the analyst can read an audience to determine their openness, defensiveness, readiness, frustration, confidence, boredom, and acceptance.

By reading the body language of the audience through its reactions and expressions, the analyst can gauge how to adjust his style and the pace of the presentation. Questions addressed to the audience or from the audience, can be used as an indicator of audience understanding.

REPORTER VS. SALESMAN

In most educational settings, it is commonly assumed that the analyst should be an unbiased, objective reporter of facts to management. My own experience however has taught me that to be a successful implementer of system change, the analyst must often be a hard-hitting salesman. The question that needs answering then is, Where on the reporter-salesman continuum should a system analyst be? (See Fig. 15-1.)

For reasons that are fundamental to human nature, it is clear that in no way can anybody be an unbiased, objective reporter of information (point A). One is always biased because he cannot see the total picture, he views events in relationship to his mental sets, etc. Man is inherently biased. If people want an analyst to limit his comments to just facts, the analyst will have almost nothing to say about future events. Most systems work is subjective in character, though hopefully

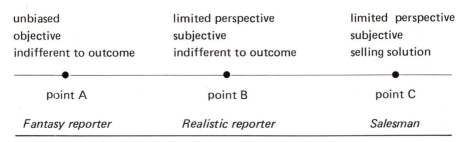

Figure 15-1 Reporter-salesman continuum.

enhanced by experiments, models, and experiences. But it is in essence still basically a matter of judgment.

This being true, an analyst often works from a limited perspective and uses information which is highly subjective. However, he still remains indifferent to which alternative system design is selected (point B). This position would still seem acceptable to most professional codes.

How then could we suggest that the system analyst should be a hard-hitting advocate out to sell a particular solution to management (point C)? The basis of this position is *relative competence.* Who has the competence to make the best decision regarding the system solution? There is no question who has the authority; that is established as the CDM. But who has the competence—the analyst, the staff, the workers, the CDM? The answer is, it depends.

To say that the CDM always is the most competent implies either a tremendous faith in top management or very simple problems. CDMs are usually very sharp individuals with good track records. However, they are also very busy people who have very limited time in which to think through problems. That's why they have analysts. The analyst, on the other hand, has just immersed himself in the study of the system and how it works. Few people will have as good an overall knowledge of the system as the analyst does. It is not a question of who is smarter. It is a question of who has had the time to take a look at the total system.

Given this as reasonably true, why throw away the expertise of the analyst by limiting him to the role of a neutral reporter? If, in fact, the analyst is the most knowledgeable about the system in question, why isn't it in the best interest of the organization for the analyst to fight to ensure that the best solution for the overall system, as he sees it, is put into effect?

Thus, the position where an analyst should be on the reporter-salesman continuum, is governed by how much competence the analyst has concerning the particular system, and the expertise on these matters of the CDM. If the analyst, because of inexperience or lack of time, has not had a chance to do a real thorough study of the system and the CDM is quite knowledgeable in this area, it is best that the analyst's position should be one of reporting information to the CDM for his review and decision. At the other extreme are those situations where the analyst is very competent in the system area and the CDM's knowledge is quite limited. Then the analyst should drive for the best system solution as he sees it. The other combinations of CDM knowledge and analyst competence would result in recommendations somewhere between point B and point C.

We note that the question of whether the analyst should be a reporter or a salesman is concerned with communicating the study results. This comes *after the study has been completed* and the various system designs have been evaluated. At that point, an analyst has a choice to make, whether to be a reporter or a salesman. During the initial analysis of the system and the evaluation of alternatives, the analyst has no choice! He must be indifferent to the outcome.

PREPARING FOR THE PRESENTATION

Whether the presentation of the final system study results are to be in a written or oral form, the analyst has to do some prior preparation to: (1) recheck the "correctness" of his analysis and conclusions, (2) sense what the general consensus of response will be to the study results, and (3) develop a presentation strategy.

Recheck "Correctness" of Study

As has been stated often in this book, to do any system study the analyst needs to make many assumptions, estimates, and just plain guesses. In the presentation of system design alternatives, many people including the CDM will be looking to the analyst as the expert on the system. Not because he is necessarily smarter than everyone else, but more because he is the only one who recently has had a chance to study the total system in some detail. The analyst thus needs to review at this point, after all the "dust has settled," what in fact are the basic conditions or specifications that are being used for the system (i.e., criteria, constraints, planning horizon, scenario, etc.). Do they still make sense in light of all the things that have been learned since the study began?

However, in rechecking the validity of the study results, the analyst can't become overly demanding. In the study of any complex system, one can't gather all information. There will always be many "holes" in the analyst's support base for his conclusions. The analyst just needs to do the best that he can and make explicit where possible the underlying assumptions, confidence level, etc.

Determining General Consensus

The analyst needs to talk (as he should have been doing throughout the study) to key people like the CDM, functional heads, and workers to get some idea of their *present* thinking on the various system designs. This approach is useful since it helps to show the key people that the analyst is interested in how they feel. Further, it gives the analyst a chance to marshal support for particular solutions and to become aware of who is going to fight various solutions if they are introduced.

Knowing who is against various proposed solutions and how strongly they feel about it, is vital intelligence for the successful systems analyst. When the analyst talks to the various key people to inform them of what he is going to tentatively discuss, it becomes quite clear who is opposed to various options. From these people the analyst needs to determine why they oppose various plans. What are their main arguments? Using these arguments the analyst can discover if in fact there is any basis to them. If there is, he may need to change his system design. If there is not, the analyst needs to build strong counter-arguments for the eventual showdown.

PRESENTATION STRATEGY

Up to this point we have discussed many of the underlying concepts of communicating the system study results. Now we need to relate these points to the earlier elaboration of the decision framework (Fig. 13-4, p. 247) to develop our presentation strategy. For convenience, the earlier decision framework is reproduced in Fig. 15-2 with the reporter vs. salesman aspect not very relevant. Additionally, key strategy points are labeled alphabetically and will be referred to in Table 15-3, on presentation strategy.

The determination of the appropriate strategy to be used by the analyst in reporting the study results, involves deciding whether he should be a reporter or a salesman. Further, if the analyst is going to assume the salesman role, he needs to be aware of the various key points in the decision process (i.e., organization versus individual player, goal congruence versus divergence, etc.). Depending on what the situation is, the analyst will have to adjust his presentation to either emphasize: (1) what's best for the organization; (2) what's best for the organization, and discuss impact on the subsystems; or (3) what's second best for the organization, and discuss impact on the subsystems.

The major difference between these strategies and the usual discussion on how to make presentations is the explicit recognition that system study results have political ramifications. While it would be preferable for everyone concerned if this were not true, changes to the status quo often have "messy" ramifications. The analyst needs to understand how to deal with this situation in an effective manner.

Reporter

The first case, point A in Table 15-3 is determining what the presentation strategy should be when the situation calls for the analyst to

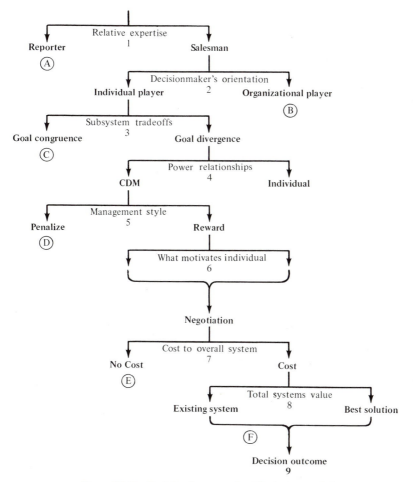

Figure 15-2 Decision framework with strategy points.

Table 15-3 Presentation strategy.

Situation	What's Best for Organization	What's Best for Organization Impact on Subsystems	What's Second Best for Organization Impact on Subsystems
Reporter	A		
Organization player	B		
Goal congruence	C		
Penalize		D	
No cost		E	
Compromise			F

be a reporter. This is the situation generally assumed in most textbook cases, and it only requires the analyst to "report the facts" of the system study results and communicate what is best for the organization.

Organization Player

A second situation, point B, is where the analyst has assumed the position of salesman, but the decision-maker's orientation is that of an organization player. Thus the analyst in effect can give the same briefing as in point A, by stressing what is best for the organization.

Goal Congruence

Point C shows a situation where the analyst is playing the salesman role, and the decision-maker's orientation is one of an individual player, but the subsystem tradeoffs are such that there is goal congruence. In this case, the analyst should present the study results in terms of what's best for the organization. The individual player should be able to sense the favorable impact on "his" subsystem generated by the overall preferred solution. If he isn't able to perceive this, the analyst could more explicitly cover the subsystem implications and why they are favorable to this person or subsystem.

Penalize

In situation D, there is basic goal divergence when dealing with an individual player. Further, the power relationship is such that the CDM has control and his management style is to penalize deviation from organizational goals. Here again, the analyst should present what is best for the organization. Additionally, there needs to be explicit discussion of the impact on the subsystem under the various options.

No Cost

Point E shows the case where the way to achieve goal congruence, when starting from a goal divergence situation with an individual player, is to reward the subsystem by sharing the system benefits. If after negotiation, it turns out that there is no substantial cost to the overall system, then the analyst can present what's best for organization and show explicitly the impact on the subsystems.

Compromise

In the last situation (point F), there has had to be a compromise on what solution is "truly" best for the overall organization. Here concessions were made in order to get the individual player to keep from vetoing any solution other than the existing system. In this case the analyst should present what is second best for organization, as opposed to what's best for organization, and explicitly show how the subsystem will benefit from this solution.

In summary, about 50 percent of the time the analyst, even when acting as a salesman, should present the system study results as he or she has normally been taught, i.e., what is best for organization with no real concern for personal impacts. This covers points A, B, and C.

However, if this straightforward strategy is what the analyst uses all the time, she will find that many times the recommended solution will not be accepted or actually implemented. Since the ultimate objective is to get better system solutions implemented, the analyst will need to have the ability to compromise on some positions and show explicitly where necessary the resulting impacts on individual subsystems.

PRESENTING STUDY RESULTS

For a presentation to be effective, we have seen that it must be attuned to the audience. When there is a large audience, the analyst is more or less forced to seek a middle position. However, when making a presentation to the CDM, for example, the analyst can specifically focus his presentation toward that person. To a somewhat lesser extent this is true of all key people the analyst interacts with, i.e., staff, workers, etc.

However, CDMs' decision styles vary significantly from very intuitive to very analytical in nature. Furthermore, there are generally many other people to whom the analyst must communicate the study results, at different times, and in different settings. Thus, for ease of explanation we are going to suggest a way to make verbal and written presentations of the system study results to composite audiences—that is, when there is more than one person in the audience and they represent the full gamut of decision styles.

Verbal Presentation

An oral presentation is a very powerful way of putting ideas across and getting decisions made. These type of presentations to executives are almost essential, where problems of major importance are being considered. This is true because higher-level executives (1) have limited time to read, (2) desire to ask questions about the study, (3) want to hear views of people in opposition to the plans, and (4) need to size up the analyst.

Executives are very busy people with much more to do than they can possibly get to. They don't have the time nor in general do they have the technical background to evaluate studies in depth. Therefore, they need a quick and accurate picture which summarizes the problem and alternative solutions.

When an executive reads a system study, he cannot readily get answers to his questions. This often leads to confusion or inefficiency in his understanding of the later parts of the study. With a briefing, the executive has a feedback loop through the system analyst. His questions can be clarified immediately, so further main points can be understood.

If an executive reads a study, he generally gets only one perspective to the problem. In a briefing with key opposition members in attendance, additional points of view can be expressed. Counterarguments can also be given. This type of advocacy situation helps to "get all the cards on the table," so the CDM can better judge which solution to choose as opposed to a sequential arrangement of talking with each decision-maker individually.

The last point is of major importance. Because of limited time, knowledge, and expertise, a full understanding of a problem area just is not possible. Because of this, a full resolution of conflicting opinions on the best system design cannot be accomplished. In spite of this, decisions must still be made on which system to select. The CDM has to rely extensively on his judgment to make these decisions. In this regard, most successful CDM's place considerable importance in how much confidence they have in the people supporting various positions.

For the analyst to have any impact on the final decision, the CDM must have confidence in him. The CDM makes a judgment about the analyst by seeing him in action. What kind of briefing does he give? How well does he answer questions? How confident is he? These thoughts are very consistent with the need to evaluate information and its source (chapter 11).

In giving briefings on the results of a system study, the general format shown in Table 15-4 is recommended. State the purpose of the study to the CDM and others for their understanding of crucial points in the final recommendations. Talk of the alternatives under final consideration and their strengths and weaknesses. At the end, the analyst should review the major points and state explicitly his recommendations. At this point he should entertain any objections from the CDM or any other people.

The analyst should be prepared to handle most objections to the study results since he has talked with many people during the study, while setting up the presentation, etc. Where possible, the alternatives to be evaluated should include the one recommended by influential people who oppose the plan, an alternative representing the conventional wisdom, the existing system solution, etc.

The analyst should openly and frankly answer any objections to the study results. There is no place for railroading results or avoiding sticky issues, even when the analyst is functioning in the salesman's role. If the analyst can't show "his" recommended solution is better, or can't reasonably answer the objections, then he doesn't have a very strong case and should question it.

In using the systematic systems approach the analyst would present the purpose of the study, give the audience enough background to understand the problem area, define the alternatives, then present the evaluation matrix. Any necessary explanation of symbols, etc., could be discussed at this point along with major constraints and criteria definitions which weren't self-evident.

The relative strengths and weaknesses for the alternatives would be discussed and the rationale behind why the recommended alternative looked best. Explicit tradeoffs would be pointed out under the proposed solution. Discussion could then follow.

Table 15-4   Verbal presentation format.

Purpose of study
Background
Present alternative solutions
Discuss strengths and weaknesses
Recommended solution
Open for discussion
Decision

In this way, the analyst could field questions, objections, etc., and relate them to the summary perspective given in the evaluation matrix. If other points came up, the analyst could discuss the preference chart, systems utility function, etc. But otherwise, these aspects wouldn't be discussed explicitly in a short oral presentation.

While the content of the briefing of a system study appears similar to the written form, the actual presentation of materials is generally very different. The two major causes for this difference in presentation are the amount of time and degree of control available. With a briefing for high-level executives, the analyst generally has less than one hour to present the study results. Sixty minutes is a very short time to get across complex concepts. In fact, it cannot be done. We would leave out details or at least be very selective about including them in a briefing.

Complementing the relatively limited amount of time available to get points across, is the advantage a good briefer has of controlling the pace of the briefing and where it is going. This control is necessary so that the briefer can take the CDM and others through a step-by-step development to the final recommendations.

For the CDM to follow this development, the briefer needs to be sure the CDM is understanding what is said at each step. This understanding can be ensured by either communicating to him in a language at his level or by providing him with background material to bring him up to the level necessary for further discussion. How well the CDM understands the development can be shown by the type of questions he asks or does not ask during and at the concluding part of the briefing. Another indicator of understanding is the expressions on people's faces and other nonverbal clues. Good briefers develop a sense of when to reiterate points or to pick up the pace.

Written Presentation

The basic points which should be covered in an effective written presentation of a system study are shown in Table 15-5.

Table 15-5 Written presentation.

Purpose of study
Scope of study
Background information
Criteria
Alternatives
Evaluation method
Conclusion/decision
Information sources

Table 15-6 Final report--written format.

I.	*Management Summary*	*VI.*	*Expected Systems Performance*
II.	*Problem Identification*		Outcomes
	Symptoms		Confidence
	Cause		Information support
	Problem		Systems simulation chart
	Scope of study	*VII.*	*Recommendation of Best Alternative*
III.	*Systems Context*		Alternative rankings
	CDM		Decision rationale
	System diagram		Tradeoffs
	Scenario	*VIII.*	*Planned System Change*
	Systems objectives		Pressure
IV.	*Feasible System Designs*		Relative advantage
	Major constraints		Goal congruence
	Final alternatives		Amount of behavioral change
V.	*Desired Systems Performance*	*IX.*	*Systems Goals*
	Criteria		Measure of effectiveness
	Preference chart		Systems goals
	Systems utility function	*X.*	*Information Sources*

The amount of discussion on each point will depend on the value of the study and the characteristics of the audience. The purpose of the study should be made clear in an explicit statement of what is hoped to be accomplished with the system. The scope of study covers factors such as who is the CDM, planning horizon, major assumptions made, and important constraints. The main purpose of the background section is to be sure the readers of the study will be able to understand what is being studied. This may include a flowchart or general description of the system. In technical presentations to nontechnical readers, some type of definition of terms is generally helpful.

What specifically are the criteria being used to determine good system performance? A discussion of the final alternatives under consideration and their strong and weak points is needed. An annotated information sources section can be very useful in showing where information was drawn from and providing an indication of its validity. What evaluation method was used to determine the worth of various alternatives? Finally, a conclusion section is needed to specify what is felt to be the best option to select.

If we needed to communicate to just one person, whose style we knew, we could (and should) specifically tailor the written presentation to their needs. But generally there are many different people the analyst needs to communicate with using this written report. All decision styles are usually represented in the Problem-Solving Group for example. Further complicating the issue is the fact that the expertise and interest of the readers also varies greatly between managers on the one hand and technicians on the other.

One way to communicate with such a diverse group of people is to write several documents. One would be for managers giving an overview perspective of the problem and what to do about it. A second document could be written covering the more technical details to be read by specialists in the various functional areas.

Another approach is to write one document which takes a composite viewpoint for a diverse group of readers. This is the method which is shown in Table 15-6, when writing the final report using the systematic systems approach.

The details for this report were covered in chapter 10. Here we want to go more into the reason behind why it is formatted the way it is. The first part of the final report is a management summary which gives an overview of the total study and what alternatives were considered, and what is recommended for the future. This summary is typically limited to one page.

Sections II through VI follow the build-up method in which the problem is identified, the systems context is defined, feasible system designs are described, desired system performance is specified and expected systems performance is validated. To help communicate these ideas quickly and effectively with the systematic systems approach, summary charts are used such as the preference chart, systems utility function, systems simulation charts, etc.

The evaluation matrix is used to display all the resulting aspects of the study and to summarize for the reader what has been accomplished, show how good it is, and suggest where more study might profitably be done. The discussion centers on the decision rationale and the resulting tradeoffs.

Next, what system factors need to be considered regarding the implementation of the recommended alternative are given, along with the resulting systems goals. The information sources used and their relative accuracy is given in the last section. Much more detailed analysis or simulation results, as an example, can be in the appendix sections.

With this type of format for the final report, the reader can approach the document in various ways. He could of course simply read the study from the first page to the last. This would give the reader a logical build-up to what the problem is and why a certain solution was recommended. For other readers, who might have much less time or motivation to read the complete study, the management summary is provided. If a manager only read this page, he would still have a good idea of what the study was all about. Certainly not much detail, but a good overview.

If the reader wanted to selectively read the report, she could read the management summary, then go to the evaluation matrix. With this one-page overview, the reader would have a very good concept of the alternatives, criteria, relative worth, tradeoffs, etc., through the evaluation matrix. If the reader didn't understand what a certain criterion meant, she could go back to section V and just read the definition for that criterion. If the person didn't agree with the ratings for an alternative on a particular criteria, she could selectively look at a specific system simulation chart in section VI and compare it with the relative performance measures shown in the systems utility function displayed in section V. Furthermore, if the reader doubted the confidence indication shown, she could refer back to section X, where the specific source is evaluated as to its overall confidence level.

In summary, the final report format shown in Table 15-6 serves multiple purposes and can be used for a composite audience. It allows both front-to-back reading and a much more selective management-by-exception approach which zeroes in on issues the reader is most concerned with.

Class Exercise 15-1
READING PEOPLE

In their book entitled *How to Read a Person Like a Book*, Nierenberg and Calero state that much can be revealed by reading a person's *total* body language. For each of the following attitudes, state what you think the nonverbal clues would generally be. That is, state how a person would sit or stand, what they would do with their hands and legs, and what expression they would have on their face.

1. (a) Boredom
   (b) Defensiveness
   (c) Acceptance
   (d) Nervousness/suspicion
   (e) Readiness
   (f) Frustration

2. Assume you were interviewing a foreman, and by reading his body language you concluded that he was very defensive and appeared suspicious of you and your questions in regard to certain delicate issues. How could you change your interviewing approach, so that you can get his real views on key points. If you can't accomplish this, what should you do?

3. Assume you were giving a briefing to an executive (CDM), and you felt he was bored. How could you adjust your style? What if he were just preoccupied with other thoughts?

Class Exercise 15-2

COMMUNICATING WITH THE CDM

Assume you are in a situation as a senior systems analyst, whose team has completed a major system study. Since the results of this study can have major implications for the future success of the organization, you want to be sure you effectively communicate the crucial points to the CDM. Therefore you are very concerned with determining the decision style of the CDM, the best mode to communicate the study results, and what your presentation strategy should be.

In talking with other people who have worked with the CDM, looking at his biographical sketch, and from observing him and his office decor, you have established the following. The man has risen rather quickly through the organization, for which he has worked during the last 20 years. His accounting background has served him well in his new position of vice-president, controller. In evaluating projects he often asks what is the "bottom line."

He is a man of action. His office, while large, is very conservatively outfitted and everything seems to have its place. According to his secretary, he meets with many people during the day and the only way of getting an interview with him is to set up an appointment.

1. What decision style do you feel best fits the vice-president, controller? What is the best method for communicating the study results to him? If he requests a verbal briefing, how would you conduct it? If he says that he has to leave town and would like to read the study while he is on the airplane, how would you meet this request?

2. After presenting the study results to the CDM (i.e., vice-president, controller) of this organization, he states that in view of the potential impact on other areas of the business, he would like you to give an oral briefing to all the vice-presidents and himself. You are very concerned since the study results have quite a negative impact on the future of the production department. Further, the vice-president in charge of production has been very cynical toward this study. You have heard through the "grapevine" that he is concerned that the study recommendation of reducing the importance of the production department will hurt his chances of advancing in the organization. As you looked at the alternative of compromising to appease the V.P. for production, you see this presents too much of an economic cost to the organization.

   (a) What type of player do you feel the V.P. for production is?

   (b) In determining a presentation strategy, what is the first question you have to ask yourself?

   (c) What presentation strategy would you use?

Chapter Sixteen

Problem-Sensing

Problem-sensing involves those methods which alert one to the fact that the system is not performing as desired and therefore change of some sort is necessary. In this chapter we will discuss the very important followup actions which need to take place after the selected system design has been implemented. The discussion will show how system goals, feedback, and monitoring are interrelated. From this background, the fundamental causes of systems problems will be developed and examples given of each basic type.

FEEDBACK

Because we are dealing with complex systems projected into a future time period, it is to be expected that things will rarely turn out exactly as forecasted by the results of our system study. Some aspects of the implemented system solution will be giving better performance than expected, whereas other features will be showing lesser performance than we thought. This is especially true for those criteria where

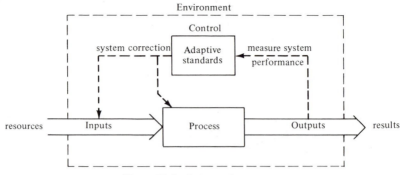

Figure 16-1 Cybernetic system.

the confidence levels are not that high (i.e., systems simulation—low or no confidence).

Thus, there is a need to consider feedback as an integral part of the systematic systems approach, to ensure more effective systems performance during the planning horizon.

> **Feedback** is the comparing of *actual* systems performance with expected performance. Based on this comparison, a readjustment is made of system factors to come closer to the desired results.

To further elaborate on this concept, recall the basic cybernetic systems diagram of chapter 3.

Various kinds of inputs such as men, materials, machines, money, and information are drawn from the environment by the system. These resources are then converted to system outputs in the form of products, services, increased knowledge, etc. To see whether the system is performing well, we must measure its performance against some specified standards. If the system is not performing as desired, system corrections are initiated to either change the system inputs or the conversion process itself to give a more effective system performance. Or in the case of a highly changeable environment, the standards themselves may have to be adjusted to a more or less demanding level. This total comparison and adjustment process is what feedback involves.

CONTROL STANDARDS

We have discussed control standards implicitly throughout the book. In Fig. 16-2, we show schematically how the various control standards relate. Initially, when we looked at a system and its overall purpose, we generated a set of system objectives. As suggested by the

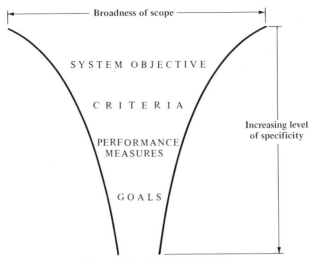

Figure 16-2 Control standards.

figure, these standards of performance are usually very broad in scope and are stated very generally. To get a better idea of what was considered effective system performance, we next went to establishing criteria. The criteria defined more specific aspects of the system objectives.

Performance measures were then developed for each criterion to show what would be considered barely acceptable, average, and exceptional performance for the type of systems under study. After the various proposed alternatives were rated on these performance measures, an alternative was selected. Since this alternative was considered the best solution overall for the system, its expected performance rating on each criterion could be used as the system goals. These goals then became the control standards to determine whether or not a system is functionally effective during the planning horizon.

DEVELOPING CONTROL STANDARDS

With the chosen solution now in place, we need to develop some type of information-gathering method to establish how well the system is in fact performing. Many large, sophisticated organizations monitor their performance through elaborate (and costly) management information systems. Other smaller systems use much more informal methods, such as reacting to crisis as the main indicator of problems.

Irrespective of the size and sophistication of the monitoring schemes, they need to be tied to the results of the system study in the following way. System effectiveness measures need to be established

293

and then target levels of performance should be specified. The results of a system study give this information.

>**System effectiveness measures** are those criteria used to monitor present and future system performance.

The criteria that have been developed to judge the various system design alternatives should be those factors which the CDM and others feel are the most important aspects involved in judging how effectively the overall system is functioning. Therefore, we find that the system effectiveness measures have already been specified in the evaluation matrix.

>**Systems goals** are the expected levels of performance which a system should be attaining during a future time period.

Since the selected alternative design has been judged to represent the best *overall* performance that can be reasonably expected of the system during the planning horizon, the resulting level of performance on each criterion will be taken as the system goals.

EXAMPLE OF CONTROL STANDARDS

Let's reconsider the example used to determine the best way for a suburban newspaper to set type. More specifically, we look at the evaluation matrix, systems simulation chart, and the systems utility function shown in chapter 12.

The criteria used for the study of typesetting systems were capital investment, time element, page cost, maintenance of system, input volume, versatility to customer, personnel competence, status of personnel, and product quality. For this particular subsystem of the overall company, these criteria will be used as system effectiveness measures.

For the systems goals, we can take the expected level of performance shown in the evaluation matrix for the selected alternative. In this case, let's assume it was hot metal process B. The resulting system goals are shown in Table 16-1.

Thus, we can expect to make a capital investment of around $353,000 with a page cost of $2.22, if the hot metal process B is implemented. Since the accounting staff generally is the monitor of dollar costs, we now have a set of figures which can be used in that portion of the future budget which is concerned with the setting of type for the newspaper. To the marketing staff, we can indicate to the

Table 16-1 Control standards for typesetting systems.

System Effectiveness Measures	System Goals
Capital investment	$353,000
Time element	32 lines per minute
	7¼ hours
Page cost	$2.22
Maintenance of system	Company machinist and service policy
	Trouble-shooting of weak areas
Input volume	6 takes
Versatility to customer	60 fonts
	Black and two colors
	Two-day salesman proof
Personnel competence	Accurate makeup
	Responsible worker
	General knowledge of process
Status of personnel	Keep all ten employees with company
Product quality	Solid type faces
	Clean, without scum
	Color balance
	Good alignment on type

salesmen what level of versatility to the customer will be available to potential advertisers with our newspaper. More specifically, this will be 60 fonts, with black and two colors. While this is not exceptional, it is better than what is available with the existing system.

As for the production department, we can let them know that for scheduling purposes, we will be able to print the Sunday edition in 7¼ hours with the new process. Furthermore, any nonstandard runs can be expected to be printed at 32 lines per minute.

To the personnel director, we can show her what the status of the employees will be under the new alternative. In turn, she will be able to forestall any strike by the labor union, since all ten employees will be retained within the company.

PHILOSOPHY OF GOAL-SETTING

Now that we have an understanding of how systems goals have been defined and how they are to be used in this methodology, it is important to contrast this approach with the more usual method of goal-setting.

Usually people speak of setting goals as the initial input to a systems study. In other words, the purpose of the systems study is to detemine how to attain the stated goals. In direct contrast, the systematic systems approach states that the systems goals are the output of the systems study. That is, the purpose of the study is to determine what are the best results that can be attained from a particular system.

Why this great difference in the approach to systems goals? Part of the difference is due to people not distinguishing between objectives, criteria, and goals. Most people use these terms interchangeably. We have developed very specific definitions for each of these terms.

As an illustration, consider the situation where the CDM really means systems goals as the desired output of a system. But in the study of any complex systems problem, how can the CDM or others know what are "reasonable" goals without studying the situation? By specifying arbitrary goals, the CDM can lead the analysts off on a "wild goose chase."

In contrast, we can have the case where the CDM does have a thorough understanding of the system and does define reasonably specific goals in the initial phase of the study. With the iterative systems approach, this "starting at the end" can be accommodated. However, the CDM and analyst should both be open-minded enough to take the initial goals as guidance, which can later be modified if the study results so indicate.

In addition to the difference in definition of what systems goals are and when they should be specified, there is the philosophy of what you are trying to accomplish with the systems goals.

Usually the CDM is trying to make systems more effective during some future time period. However, the difference comes about in how this message should be conveyed to the people who are a part of the system. Many managers feel that one should set goals which are unrealistically high (i.e., "aiming for the stars"), even though they know this isn't attainable. They feel people will work harder and thus accomplish more this way than if you set "realistic" goals. The most desirable performance would be given by taking what has been specified as exceptional or outstanding performance in the systems utility function for each criterion. However, because of the concept of system tradeoffs, we know this is not really attainable.

Setting goals by this latter means would (1) lead to frustration on the part of everyone who would try to attain that which is not attainable and equally, if not more important would (2) send misleading signals to analysts and managers who would be going through the costly procedures of changing and adjusting inputs, outputs, and processes needlessly.

The approach taken in this text is the philosophy that people will respond more effectively if you give them realistic goals which are the result of a study. Therefore, if the systems goals were taken from the evaluation matrix as the expected performance for the selected alternative the level of performance is not only attainable, but has been judged to be the most reasonable under the circumstances. Further, it can be noted that if these goals are not met in the future, there is something wrong with the system which, in fact, needs changing.

On the other hand, in setting the "aim for the stars" type of goals, since you know people can't attain these goals how do you reward or penalize performance?

MONITORING SYSTEMS PERFORMANCE

Now that we have developed a set of control standards and more specifically, systems goals, these can be used to monitor and assess systems performance. There will be three major times when we will want to do this, starting with *acceptance testing,* then *systems audit*, and finally as a means for *sensing a new problem.*

Acceptance Testing: In the development of alternative solutions as discussed in chapter 5, it was indicated that frequently the alternatives were already developed (i.e., off-the-shelf, existing system, etc.) and thus could be tested prior to a decision being made. Oftentimes, however, the alternatives evaluated in a systems study are being proposed and thus haven't been completely developed (e.g., new systems, extensively modified existing system).

If one of these latter alternatives has been selected as the best solution, then the analyst and other appropriate organizational people need to develop a means for accepting or rejecting the newly designed alternative. That is, when an alternative is to be designed to certain specifications, how does the CDM ensure that it meets all these performance requirements?

The analyst can many times directly use the systems goals as the means for acceptance testing. For example, in the development of a computer system for a small business, it might have been determined from the systems study that the time to complete the monthly accounting statements was to be three hours of computer run time. Further, the whole process was to be handled by an accounting clerk with no computer knowledge working at a terminal (i.e., TV screen). Lastly, the programs were to be run on the $25,000 computer system that the salesman had suggested.

With these specific goals, the analyst can now determine whether the targets had been met, and thus whether to accept the solution. When possible, this type of process should be included in a contract with appropriate penalties for nonperformance.

Systems Audit: This is a followup study which is concerned with the whole system as it was affected by the results of the systems study. That is, after the study has been completed and implemented, and the system has been operating in this new mode for some amount of time (two months, one year, etc.), an audit is done to see how well things really went.

Here again, the systems goals fit in perfectly. The expected future performance of the new system solution has been defined in very specific terms. Further, these are realistic goals of what should be attained and they have been agreed to by the problem solving group (i.e., CDM, decision-makers, analysts, etc.).

The function of the audit is to see how the actual system performance is on all dimensions and to establish where deviations occur. Where there are discrepancies in performance between what was expected and what is actually taking place, the audit team tries to discover what has caused the difference. That is, was the problem-solving group too optimistic, were there poor acceptance testing procedures, were there significant changes in the environment not mirrored in the assumed scenario, etc.

The purpose of the systems audit is not only to try to improve the present situation, but it can be used as a learning device to both perform better system studies and to better educate all concerned.

Sensing a New Problem: This facet of monitoring systems is concerned with assessing the on-going performance as it related to the expected performance (i.e., systems goals). It can be done on a management-by-exception basis, bringing attention to management only when there are relevant discrepancies. For example, assume that coming out of the system study was the systems goal of accepting a 5% rejection of products by the quality control division of a manufacturing firm. Appropriate tolerance limits would have to be set so that management isn't overwhelmed with transient fluctuations but rather is told of significant deviations.

In Fig. 16-3, this management-by-exception approach to monitor-

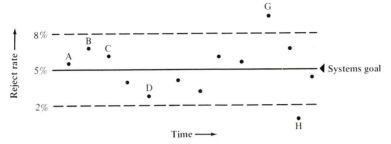

Figure 16-3 Monitoring systems performance.

ing systems performance is shown with a systems goal for product rejections set at 5%. The rate shouldn't be arbitrary, but rather should be the output of the systems study.

The tolerance is set at ±3%. In other words, if when the quality control people monitor the production line, as long as the rejection rate over a period of time stays between 2% and 8% with an average of 5%, we will assume there are not significant problems since the systems goal is being met.

Point A, point B, point C, and point D in the figure are all within limits and are considered statistically acceptable fluctuations. Point G and point H are outside the tolerable limits and therefore potential problem indicators.

With point G, the rejection rate is 10%, which is too high. Management should be advised and appropriate action taken. But why would Point H be considered a problem since it shows a favorable condition of a low rejection rate of 1%? Since complex systems are interrelated, it might indicate that the quality control division has become too lax and months later there will be a rash of consumer complaints. Or it might be that the production line workers are trying to do too good a job, and thus expending too much time on each product.

In any case, this analysis suggests how the systems goals can be used to sense when a new problem has developed, by monitoring the various indicators or symptoms of systems performance. This now brings us full circle in the systems approach to problem-solving.

The symptoms help us to formulate the problem. From there, we gather information in developing system solutions, etc. Relook at Fig. 1-1.

Problem-solving is a never-ending process. The system doesn't stay solved even though the solution was effective. The environment could have changed or maybe a new CDM takes over, etc. We need to appreciate that any new problem is different from the original one; i.e., time has passed and conditions are different. In turn, this leads us to the discussion of the fundamental causes of problems.

FUNDAMENTAL CAUSES OF PROBLEMS

From the establishment of goals, feedback methods, and monitoring comes the recognition or at least the possibility of a need for change with particular systems. The analyst and/or manager will determine that these indicators are valid and that they do point to basic problems with the system.

For the analyst to be effective at this point in ascertaining what the problem really is and determining what kind and how much effort should be put forth in arriving at a solution, it will be helpful to be able to differentiate among the fundamental causes of problems. As we previously defined them, problems are those situations in which the present or future performance of a system is no longer acceptable or viable.

In Table 4-1, we classified the types of problems to be negative, positive, and fundamental deviations between what was expected and what was actually happening. Since chapter 4, many more concepts have been defined such as constraints, objectives, benefits, costs, etc. Now we want to use these additional concepts to see more specifically what the fundamental causes of problems are.

> **Fundamental causes of problems** are the primary factors which are creating the difference between expected and actual system performance.

As shown in Table 16-2, there are basically three main causes of problems. The first type of cause arises when some aspect of the system is violating some particular constraint limit, thus making the system no longer viable. This condition could be brought about by some component of the system deteriorating or failing. Or the system could be functioning as it was designed to do, but now some constraint limits

Table 16-2 Fundamental causes of problems.

1. System is infeasible	2. Better solutions	3. Change of system objectives
(a) Component deteriorated or failed	(a) Can accomplish same benefits for less cost	(a) Higher aspirations
(b) Constraint limit changed (i) Internal (ii) External	(b) For same costs can give increased benefits	(b) Lower aspirations
	(c) Improved benefit/cost ratio	

have been changed. These changes could have been generated internally under control of the CDM or by environmental changes which result in different external requirements.

A second situation involves the system that is viable and operating as it was designed to and is therefore accomplishing the stated objectives. So, in effect, there are no problems. However, when a new alternative solution is presented to the CDM to enable the system to more efficiently accomplish the stated objectives, the second fundamental cause of problems is encountered. Here, because of new technology or access to more talented workers, etc., the system can (1) attain the same benefits for less cost than the system presently has or (2) for the same present system costs, increased benefits can be attained or (3) by expending greater amounts of resources (i.e., dollar costs), the system can attain correspondingly greater benefits.

The third fundamental cause of problems occurs when the present system is viable as it stands, but the system objectives are changed. Objectives can become more demanding simply by raising performance requirements; or the standards can be effectively lowered as in the case where the system solution is too good for what is really required (i.e., overkill). Results from a system trying to better adapt to its environmental context.

Examples of Problem Causes

To show how fundamental causes of problems relate to systems, different organizational systems will be considered below as examples for each of the various causes.

System Is Infeasible: The Lockheed Aircraft Company can be looked at as a system for producing commercial and military aircraft. In the course of building many aircraft, Lockheed has had need to borrow substantial sums of money. The banks which have lent this money put various restrictions on Lockheed to protect their investment. Assume one of the constraints is that Lockheed must have a minimum of one million dollar cash balance at all times. If ever Lockheed should have cash flow problems such that the cash on hand drops below the minimum stated, this business system is no longer viable. Lockheed now has the problem of getting the cash flow within acceptable limits, while pursuing its unchanged system objective of profitable building of aircraft.

In the above example the constraint limit was constant, and the system operated outside it. A second case is when the system is functioning well, but the constraint limit has been changed which causes a problem. The situation of a changing external constraint is easy to see in relationship to say governmental intervention. The military services of this country can be looked at as organizational systems with the overall objective of providing military force when needed. A major input to the military service is young men. When Congress decided that the draft system was no longer acceptable, this presented the military system with a problem. The changed external constraint can be stated as: The military must get its manpower by voluntary means. To keep the system viable under these new conditions requires the military to drastically change its recruiting policy and the way it treats men while they are in the service in hopes of alleviating future retention problems.

Better Solutions: As was shown in chapter 3, we can look at any business as a system which has some particular objective like maximizing profits. If a system is meeting all internal and external constraints and making progress toward its system objectives, it is in fact viable and there are no problems. However, the fact that there are no problems does not imply that the system cannot be operated more efficiently. An example of this occurred when commercial computers first became available in the 1960s. Many an IBM salesman created problems for organizations. The mission of these salesmen was to show executives how, by computerizing the present manual accounting system (i.e., payroll, accounts receivable, etc.), the organization could decrease operating costs while keeping the same benefits. This forced company executives to face the problem of determining which method (manual or computerized) best helped the organization meet its overall objectives.

Sometimes improvements can be obtained with no change in operating costs. During the late 1960s and early 1970s the management of pension funds became increasingly the province of management investment firms. Previously, for most firms, pension funds were handled by the trust departments of major banks. The management investment firms were able to show various organizations that, for the same costs they were presently paying to have their pension funds managed by the banks, the investment firms would, through more astute management, give these client organizations a much greater return on their money.

Frequently a major system improvement requires significant expenditures. During the 1970s there was a technological breakthrough in the steel industry. The development of the basic oxygen furnaces for making steel has caused a problem for various steel-makers. The new steel-making technology is much more efficient than the present open-hearth furnace but it requires a larger scale of operations. To get the increased benefits of the new process, a company may have to move to a larger, more costly, scale of operation which will require a major investment in new plant and equipment now. In other words, to get the increased benefits the steel firms will have to accept more costly operations. The problem thus revolves around deciding whether or not the additional investment is worth it. That is, will in fact the benefit/ cost ratio increase to enable the company to meet the system objectives more efficiently?

Change of System Objectives: Problems can arise for an organization when it is decided that the present system objectives are no longer appropriate. For example, assume you were the president of a company like Memorex during the late 1960s. Memorex at that point was the maker of computer peripherals and, while it was quite successful, Memorex was only a small factor in the data processing industry. One morning you get a call from an RCA executive. He states that RCA is thinking of dropping out of the computer business and they are giving several companies two weeks in which to make a bid for RCA's present business. Are you interested? You now have a systems problem. You need to decide whether Memorex should stay with the present, small but quite successful peripherals business or grab at a very unusual chance to become a major factor in the much larger and more risky computer business.

On the other hand, a different type of problem develops when the aspirations for an organizational system are lower significantly because of changed environmental conditions. Take the case of a firm like Penn Central. This organization has been operating passenger and freight railroad business in the northeastern section of the country for some one hundred years. However, by the 1970s many of the industries which were the mainstay of the Penn Central freight business had either decreased production drastically or moved to the South (outside of Penn Central's territory). Passenger business for railroads had been steadily decreasing since World War II. Therefore, even though the management of the Penn Central organization system would have liked to keep the same level of system objectives they had in the 1930s and 1940s, the world had changed. If there was any hope for this system to remain viable, the management had to decrease the scope of Penn Central's operations and lower the system objectives to adapt the system to its new environment.

Table 16-3 gives an overall summary of the relationships between fundamental causes, type of problems, decision levels, and systems study purpose.

RECOGNIZING THE NEED FOR CHANGE

Any system needs to have a means of determining when it, in fact, has problems and a means for establishing what measure of

Table 16-3 Summary of problem relationships.

Fundamental Causes	Type of Problem	Decision Level	Systems Study Purpose
System infeasible	Negative deviation	Operational	Viability
Better solutions	Positive deviation	Tactical	Efficiency
Changed system objectives	Fundamental deviation	Strategic	Effectiveness

response or change is required. Before any system can decide whether it is doing well or is in deep trouble, a determination must be made of what the objectives are or should be for that system. When these objectives are spelled out in some detail, the system managers have a means for determining what is good and bad performance.

Some kind of information system must be developed to indicate or measure how well the system objectives are presently being met, or will be met in the near future. With appropriate tolerance levels around each goal, a management-by-exception policy can be initiated whenever the system gets outside of these limits. This monitoring scheme alerts management that the system is not performing as desired, and therefore a change of some sort is necessary. A determination of what the fundamental cause of the problem is will give clues as to what specific action to take.

Examples of the interrelationship of goals, feedback, and monitoring schemes of several systems are given below.

Power generating plant—electronic gauges sense the load on various generators. If the load exceeds certain limits, the system either shuts itself down, turns on a red light to alert an operator of the system condition, or automatically adds a reserve generator to the system to handle the overload.

Human body—the temperature system of the body is an example of a feedback system which attempts to maintain the system goal of 98.6 degrees. If the body goes outside these limits, sweating or shivering takes place automatically to bring the system temperature back into tolerance. If the body cannot do this, a fever develops of say 104 degrees and the person is in fact sick. The person now goes to a doctor to get a more powerful means (e.g., antibiotics) to eliminate the cause of the problem and thus restore the body into a healthy condition—that is, to get the system back within temperature tolerance limits.

U.S. Postal Service—In the late 1970s the expectation of Congress in making the United States Postal Service a quasi-independent organization, was that this reorganization plan combined with the very substantial postal rate increase, would enable the postal system to run at a slight profit while concurrently increasing the level of mail service to the public.

Soon after the plan was implemented, a subcommittee of Congress monitored the actual results and then compared this with the original expectations. Feedback to Congress was provided by the public, who collectively wrote in overwhelming numbers stating how dissatisfied they still were with the mail service. Additionally, the Postal Service itself had to go to Congress and ask for several billion dollars in emergency funds so they could meet their payroll for the year.

These indications or symptoms told Congress that there are some fundamental causes to the postal problem which need to be faced, if Congress hopes to make the mail service efficient.

Persons—Individuals need to monitor their world to see how they are doing in relationship to what can be reasonably expected. This analysis can take the form of physical and mental indications of stress like headaches, ulcers, being continually uptight, or even a heart attack.

In terms of the suitability a person has for the career he has chosen, indicators could be whether one has been passed over for promotion or given a series of progressively more responsible jobs in the organization, how much one likes to go to work as opposed to longing for holidays and vacation, or the indications could be in terms of how much money one had expected to make after five years with the firm versus what he is actually making, etc.

These examples plus many other indicators can be used by an individual as feedback to tell how things are going. They are a form of reality-testing which people, who are functioning as cybernetic systems, could use to guide themselves toward a "happy and productive" life.

Class Exercise 16-1

SETTING GOALS

Assume you were a systems analyst for the president of a small state college. The president was very interested in getting more recognition for the college and he had chosen as one of the primary methods, the development of an outstanding football team. The president reasoned that the sports coverage of an outstanding team by the newspapers, sport magazines, and possible television coverage would gain the college quicker recognition than the much slower press for academic excellence. So the president selected a football coach with an outstanding winning record at the community college level.

1. The football coach, in his speech to the student body, stated that he would make the college football team the finest in the country. As an analyst, what does the coach's statement mean as a systems objective?

2. Given that the state college team hadn't had a winning season in its thirty-year history, what would you set as a systems goal? How would this

systems goal differ from that for a new coach of the Notre Dame football program? Why the difference?

3. The small college team won 3 and lost 7 the first year of the new coach, won 5 and lost 5 the second year, and the third year won 7 and lost 3 and came in second in the league. The coach was fired that third season because they hadn't won the championship and certainly weren't the finest football team in the land. Discuss the possible differences in the philosophy of setting goals between the president, the coach, and the alumni.

4. How do you set workable goals? Is winning the only reasonable criterion for assessing a football program?

Class Exercise 16-2

DETERMINING CAUSES OF PROBLEMS

In the following examples, determine what is the fundamental cause(s) of each problem. Relate your analysis to Table 16-2.

1. During the late 1970s and early 1980s, the Chrysler Corporation went through a major rethinking of their car business. They had lost a considerable share of the market in auto sales. They had to ask for the Federal government to provide loan guarantees. What type of problem did they have? What was the fundamental cause(s)? What problem-solving strategies did they use? Were they acting as a cybernetic, homeostatic, or dynamic system?

2. Two young men, who dropped out of college, started the Apple Computer Company in their garage. Various versions of their revolutionary idea of a microcomputer for personal use were very successful. So successful in fact, that they had to come out with a version for use by small business, set up a retail distribution system, etc. To expand as rapidly as was needed, they had to raise capital by going to the investment market with a public stock offering. What is their problem? What was the fundamental cause? What future problems can they anticipate?

3. As a college student in the last part of your junior year, you attend a private college. You commute each day, 60 miles round trip from your parents' home where you stay. While you don't have a lot of extra money, your part-time job provides enough income to get gas for your car, money for tuition and books, and limited social life. You are on schedule to graduate in one more year by taking 17 units per quarter. Then OPEC causes another oil embargo and gas becomes rationed by the Federal government to 30 gallons per person per month. Further, this gasoline costs $3.00 per gallon. What type of problem do you have? What is the cause? What are some potential alternative solutions?

Chapter Seventeen

Overview
of the Systematic
Systems Approach

In chapter 1 we stated that the systematic systems approach was a structured way of solving complex systems problems. Further, this method required an understanding of both systems analysis and decision-making to be effective in developing and implementing solutions to systems problems. Initially, we said the systems approach included the following steps. Define the problem, gather and evaluation information, describe workable solutions, evaluate benefits and costs of solutions, recommend the best solution, present the solution, plan implementation strategy, and establish performance standards.

In subsequent chapters we have expanded on each of these points. In most cases there was a full chapter detailing each major step of the overall approach. By taking a top-down approach in the text, we were able to integrate the resulting detail into the larger framework.

We are now at a point where we can present an overall summary of the systematic systems approach which contains the concepts which have been developed throughout the text. After this has been done, an example will be used to clarify and amplify the overview chart (Fig. 17-1).

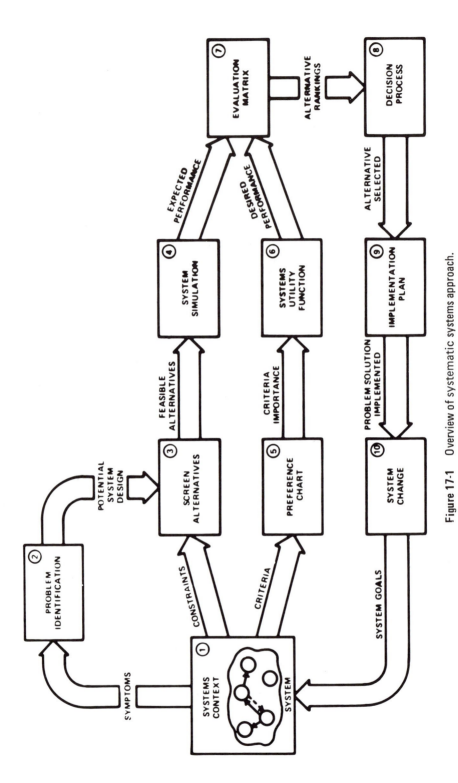

Figure 17-1 Overview of systematic systems approach.

MAJOR STEPS

As has been developed in previous chapters, the fundamental premises of the systematic systems approach are that systems should be looked at as being cybernetic, that the study of how to make these types of systems more effective requires the interrelationship of systems analysis and decision-making and thirdly, the selection process determining the best of the possible alternative designs for these systems must be able to handle complex, multifaceted solutions.

These three considerations are interwoven in the following chart which shows an overview of the major steps taken to make a system more effective using the systematic systems approach.

The purpose of each step is given as follows:

Step 1: **Systems Context.** The analyst needs to study the system itself to identify what the relevant variables are within the system and its environment. From this initial look, should come a verification of the *symptoms* of the problem. Additionally, requirements which any potential solutions must meet (i.e., *constraints*) and objectives that new system solutions should be designed to attain (i.e., *criteria*) will need to be specified.

Step 2: **Problem Identification.** From the study of the system and a review of the corresponding *symptoms,* the analyst needs to establish specifically what the problem is that needs to be solved. The analyst then proceeds to originate plausible solutions (i.e. *potential system designs*) to the system problem.

Step 3: **Screen Alternatives.** The *potential system designs* are checked to see if they will satisfy all the constraints required of the system itself. Those designs which don't are either dropped from further consideration or are modified so that they will meet all requirements. Systems designs which meet the constraints are called *feasible alternatives.* Due to the time and cost involved in designing and evaluating alternatives, the number of feasible alternatives to be evaluated in detail is generally much less than the number theoretically possible, but always more than one.

Step 4: **Systems Simulation.** For each of the *feasible alternatives,* the future level of systems performance needs to be established. That is, if each of the system designs were implemented, what would the performance of the system actually be? Since a future time period is being considered, the *expected performance* of the system cannot be known with certainty. Rather it can only be simulated, and this in turn implies varying degrees of accuracy in estimation.

Step 5: **Preference Chart.** Multiple *criteria* will need to be developed in the study of any complex system. These criteria will have varying degrees of influence in substantiating the worth of all system designs. The preference chart is a means of determining the relative *importance* of each of the *criteria* to the overall system.

Step 6: **Systems Utility Function.** In addition to the *importance* of *the criteria* themselves, the relative value of different levels of performance on each criterion needs to be confirmed. This is done by establishing what barely acceptable, below average, average, above average, and exceptional performance levels are for each of the critical criteria. The systems utility function combines the importance of criteria with the relative value of performance, to specify what is considered *desired performance* for the system.

Step 7: **Evaluation Matrix.** For each alternative design, the expected effect of implementing that solution is estimated according to a multidimensional scheme. Concurrently, a systems utility function is developed to identify the desirability of different levels of performance on each criteria. The evaluation matrix gives a means for a complex, multidimensional comparison of the *expected performance* of each alternative with the *desired systems performance.* Decision-makers and analysts are then able through this comparison to determine the *relative ranking* of the *alternatives* by considering the overall value of each system design and the explicit tradeoffs and risks involved with each choice.

Step 8: **Decision Process.** The evaluative and summary perspective provided by the evaluation matrix should be used as an aid to the decision-maker, not as a mechanical selection process. Therefore, the decision-maker, analyst, and others now need to consider in addition to the *relative ranking of the alternatives,* the risks involved with each choice in regard to future unexpected changes in the system's environment, the likelihood of successfully implementing each alternative, the need for compromise and accommodation, etc. From this process, will come the *selection of the best alternative.*

Step 9: **Implementation Plan.** Explicit consideration needs to be given on how to implement the *selected alternative.* Planning the *implementation of the problem solution* will help ensure that this system design does, in fact, get implemented in such a way that the expected performance will be realized.

Step 10: **System Change.** When the *problem solution* is *implemented,* the system itself undergoes change. The expected level of performance for the changed system, will be designated as *systems goals.* These will be derived from the expected level of performance of the selected system design.

Step 11: **Problem-Sensing.** Because the environment and the systems themselves both change over time, new problems will surely develop. To help ensure early recognition of these conditions in the future, a monitoring of actual versus expected performance should be made. Resulting deviations will be considered as symptoms that a new problem has developed. This will lead to a study of the systems context and a specific identification of the problem. Thus, the cycle is started again.

OVERVIEW EXAMPLE

To summarize the major steps of the systematic systems approach, we will discuss the selection of a place to live for the Baker family. This example has the degree of complexity necessary to show the various facets of the systems approach, yet is familiar enough so everyone can relate to it.

Mr. Baker has just received a promotion with his company, which will require him to move to the Los Angeles area from out of state. In studying the *systems context* (step 1) of the Baker family, we note that the family is composed of the husband, wife, and two children. The son is eight years old and the daughter is six. Mr. Baker's job is with a large Fortune 500 firm, where he is a middle-level manager making $34,000 per year. The Bakers have $10,000 in their savings account. They are presently renting their apartment in the Midwest. The health of the family is excellent. The Bakers are a family-oriented group, who like to spend the weekends together. Mr. and Mrs. Baker are the CDMs for major family decisions.

From the study of the systems context, it is relatively clear that this promotion is a major decision concerning the future of the Baker family. Since it involves moving from the Midwest to Los Angeles, it means finding a new place to live, changing schools for the children, leaving the relatives from both families in the Midwest, etc. However, Mr. Baker is staying with the same company, his promotion is from manager of production to V.P. of production and thus is in the same field of expertise. Overall, this is a tactical problem.

Mr. and Mrs. Baker have already completed a preliminary system study, which compared the advantages and disadvantages with accepting the promotion versus staying with the existing system. They concluded they will accept the new position. Now they need to do a more detailed study on where to live.

The problem identification (step 2) is then established as, What is the best place to live in the Los Angeles area for the Baker family? The *cause* of the problem is the acceptance of the promotion for Mr. Baker which requires a major move and makes the existing system (i.e., an apartment in the Midwest) infeasible. Mr. Baker feels that the next promotion should come in about four years and would require a move to another part of the country. Thus, the *planning horizon* is set as 1982 to 1986. The Baker family is a *cybernetic system,* which is impacted by its environment. Mr. Baker has developed a *scenario* for the 1982-1986 planning horizon. One of the important factors is the future well-being of the company he works for. Mr. Baker has concluded that the firm will continue to be very successful and it will not be a target for potential takeover acquisition.

From the study of the family and its situation, certain *constraints* can be established. For example, given Mr. Baker's future income of $34,000 per year from his new job and the availability of say $7,000 for downpayment, the maximum mortgage loan he would be eligible for would be $100,000. Another restriction on potential system designs is that Mr. Baker emphatically states he does not want to spend more than two hours per day commuting to and from work. Understanding the Los Angeles area and its lack of public transportation, and the fact that Mr. Baker's company is located in the downtown area, translates this commuting requirement into the constraint that the Bakers must live within 25 miles of downtown Los Angeles. Additionally, because of the Bakers' upbringing and their strong spiritual desires, Mrs. Baker insists that their children go to Catholic schools.

What are the *potential systems designs* for this system? They are places to live such as apartments, houses, townhouses, etc., financed by renting, buying, leasing in the neighborhoods of city, suburbia, beach area, etc. The number of possible combinations involving all these aspects is enormous. However, to greatly reduce these possibilities the Bakers have screened the alternatives (step 3) by imposing the constraints on any potential system solutions. For this family *possible feasible alternatives* must be under $100,000 total price, within 25 miles of downtown Los Angeles, and have a Catholic school within walking distance.

Remember that the best alternative can't be decided without studying the systems context in detail. In this case, the system is the particular family we are trying to help. What is best for the Bakers is not necessarily what would be best for their neighbors, or someone else. Different alternatives are best for different systems.

Assume for the Bakers it has been established that the following three system designs meet all constraints and they are the feasible alternatives.

Alternative A: A $99,500 townhouse which is five years old. It has a Spanish design with over 1600 square feet. It has all three bedrooms upstairs and a comfortable living area on the first floor. This suburban area has a pleasant climate in the winter, but heavy smog in the summer combines with many 100° days to form a rather unhealthy environment. The neighborhood is well established and a good Catholic school is nearby.

Alternative B: A classic, 35-year-old house selling for $88,000. This house is located in the inner city which is part of a large redevelopment project. The neighborhood is in flux, with the crime rate fairly high. However, it is very close to the culture centers of the city. The house is large with 1800 square feet and five bedrooms. The Catholic school here has the reputation of being one of the best in the city. The climate is moderate, with hot temperatures in summer, but generally light smog.

Alternative C: A brand new $95,000 condominium, one mile from the beach area. It is a very tasteful complex with a very small 1200 square foot, three bedroom apartment. The Catholic school is just being completed. Temperature is moderate all year around. No smog. Because of its distance from Los Angeles and the very dense traffic along those freeways, the drive to and from work will take a bumper-to-bumper trip of two hours.

Further background study of this family and what it is trying to accomplish, presently and in the future, would lead to some overall *system objectives* like growth, happiness, security, health, independence, etc. These objectives then need to be related to the problem at hand, which is finding a place for the Bakers to live. The resulting refinement leads to *criteria* which in this example are costs of the house, aesthetics of the house itself, type of neighborhood, children's education, environment of the area, potential price appreciation, house size, and convenience to work and shopping.

After much discussion about what is really important to the Bakers as a family, assume a *preference chart* (step 5) has been developed in which the *criteria* have been grouped by *relative importance* as follows: neighborhood, children's education, and environment are the most important, costs and resale value are important, and least important are convenience, aesthetics, and size of house.

In addition to the relative importance of the various criteria, the desirability of different levels of performance on each criteria needs to be determined. This is where exceptional, average, and barely acceptable performance is specified. For example, in the Baker's system, the children's education was judged as very important relative to the other criteria. What needs to be spelled out is what are the different levels of education that children can receive today at the elementary school level. By talking to parents, teachers, principals, etc., one needs to establish what is exceptional, average, and barely acceptable quality levels for education programs.

For the criterion, size of house, the question that needs to be asked is, What size place for living will $100,000 buy in Los Angeles in 1982, in areas this family considers acceptable? A real estate agent has information available which could provide an answer to the question. By considering many houses, townhouses, etc., say we determine that a 1500-square-foot dwelling is what could be normally expected. An 1800-square-foot house is exceptionally large, whereas a 1250-square-foot structure is quite small and barely acceptable.

By specifying the performance levels for each criterion, and then combining that information with the relative importance of the criteria, we have a *systems utility function* (step 6). This function will be used to judge the desirability of the alternatives to be evaluated.

Therefore, what is now necessary is to determine the future *expected performance* of the *system,* if each alternative (A, B, or C) were actually implemented. That is, how will the Baker family fare, if they were to live in each place?

This estimation is made specifically in terms of the evaluation criteria. More specifically, for each alternative dwelling: determine what the costs are in terms of total price of house, monthly payments, tax rate; determine what education programs are available at the schools the Baker children will attend; determine what the environment is like for the area; determine what the neighborhood is like now and what changes are anticipated, etc.

Some of these estimates of *expected level* of *performance* will be highly accurate, whereas others will be just guesses. Explicit recogni tion of this difference can be shown by specifying different levels of confidence in the systems simulation chart. For example, on the very important criterion, education, the Bakers have both visited the nearby Catholic schools.

The question that now needs to be answered is, Will the suburban townhouse, the inner city house, or the beach condominium be best for the Baker family? The *evaluation matrix* (step 7) gives us a way to answer this complex, multifaceted question.

With the townhouse (alternative A), the overall estimation is that the children will get a solid foundation in the religious aspects of life, good overall depth in "reading, writing and arithmetic," a fair look at the cultural aspects, and a weak physical conditioning program. The Bakers are confident of this estimation since they visited the school, talked to the vice-principal, and observed the school children twice.

The Catholic school near the inner city (alternative B) would give the Baker children a solid religious foundation, good fundamental skills in "reading, writing, and arithmetic," an average cultural background, and an excellent physical conditioning program. The Baker's are very confident of this estimation since they visited the school, talked to three teachers, talked to good friends whose children go to the school, etc. With alternative C, the Bakers' estimate is with no confidence since the school hasn't been completed yet and further, this school is going to start some experimental programs which have no track record.

In the case of the Baker family problem, the *evaluation matrix* (step 7) would look as shown in Fig. 17-2.

For each of the places to live, the type of neighborhood, the

Alternatives

Criteria	A Townhouse	B House	C Condominium
Neighborhood			
Education			
Environment			
Costs			
Resale value			
Convenience			
Aesthetics			
Size of house			
Overall Value			

Figure 17-2 Evaluation matrix for Baker family.

Criteria	A Townhouse	B House	C Condominium
Neighborhood	7	2	6
Education	5	9	3
Environment	1	5	10
Costs	2	8	4
Resale value	8	4	8
Convenience	5	9	1
Aesthetics	6	2	3
Size of house	5	10	0
Overall value	39	49	35

Figure 17-3

quality of the Baker children's education, the costs, etc., have been estimated through the systems simulation chart. These particular levels of performance for each alternative then need to be compared to the *systems utility function.* This results in a specific rating for each alternative on each criterion with 10 points representing exceptional performance, 5 points representing average performance, and 1 point representing barely acceptable performance.

Assume this has been done, and results shown in the evaluation matrix (Fig. 17-3).

From the analysis in Fig. 17-3, it appears that the house in the inner city (alternative B) is the best for the Bakers. However, this result *assumes* that each criterion is equally important and that all information gathered is complete and accurate.

Assume through the *preference chart* we weight the relative influence of the criteria and that the most important are weighted 4, the important 2, and the least important 1. The evaluation matrix would now look as shown in Fig. 17-4.

By including the weighting of the criteria, we changed the *relative rankings* of the desirability of the townhouse and condominium, but the inner-city house continues to have the greatest overall value. However, we have still assumed the information presented is totally accurate. What would happen if we discounted in some way the scores by the level of confidence we could support? This is an especially important point, since we realize that we are estimating expected system performance for a future time period. This we did in chapter 12, through the confidence levels, and developed the discounted value.

Weight	Criteria	A Townhouse		B House		C Condominium	
		R	U	R	U	R	U
	Neighborhood	7	28	2	8	6	24
4	Education	5	20	9	36	3	12
	Environment	1	4	5	20	10	40
2	Cost	2	4	8	16	4	8
	Resale value	8	16	4	8	8	16
	Convenience	5	5	9	9	1	1
1	Aesthetics	6	6	2	2	3	3
	Size of house	5	5	10	10	0	0
	Total Value		88		109		104

Alternatives

R = Relative rating   U = Utility value

Figure 17-4 Evaluation matrix for Baker family.

But for now, consider what would happen to the overall standings of the alternatives if it was determined that the funding of the Catholic school in the inner city was going to be greatly reduced in the coming year, and this meant a curtailment of many of their educational programs. Assume the net effect of this change meant the quality of their children's education would drop from an exceptional level 9, to an above average level of 7. How many points would alternative B drop? It would decrease by 8 ($9 - 7 = 2$, times a weight of 4). Subtracting 8 points from 109, leaves a net overall scores of 101. Thus the condominium now looks most attractive.

In the decision process (step 8), the Bakers need to consider the overall information supplied by the *evaluation matrix* and also check the effect of other considerations. For example, how might property tax rates change in the future among the alternatives? How would the children's education be affected if they were sent to public schools in the respective areas? How hard will it be to get a mortgage loan for each alternative and what would the consequences be of any delay? What is the possibility of a detrimental change in zoning for each of the properties? Because an endless series of questions can be asked, only the most relevant and serious ones (i.e., those with the most potential impact on the decision) can be considered further before a decision must be reached.

Assume from this process that the condominium (alternative C) looks best. Now the CDMs have together decided which alternative to select. Since it is a family and Mr. and Mrs. Baker have been playing both the role of analyst and decision-maker, there is no "chain of command" problem to contend with and minimal communication difficulties. Further, since Mr. and Mrs. Baker are both very family-oriented they are assumed to be primarily organizational players. However, Mr. Baker is an avid tennis player, and he has reasoned that if they lived in the inner city, he could play tennis an hour each work day instead of commuting the two hours to the beach condominium. As an individual player he would like that better. While this was not directly considered in the evaluation matrix, it can be covered as an additional factor.

Mr. and Mrs. Baker decide that with all the major factors now considered, the condominium is slightly more attractive than the inner-city house (104 vs. 101 points). They feel if they could get the price of the house down $5,000, it would then be best. So they enter into *negotiations* with the owner of the inner-city house, and state that they will buy it from him if he will sell it for $83,000. This price would result in an exceptional value of 10 points for alternative B on cost. In turn this would increase the total value by 4 points (2 times 2 point increase—from a rating of 8 to a rating of 10). Thus alternative B would be 101 + 4 = 105, whereas alternative C would be 104. Be sure to note that one shouldn't dissect the evaluation matrix numbers this finely; this was used just as an example to illustrate various points.

The owner states he is willing to *compromise* and bring the price down to $84,000. Since the counter-offer is now very close to what they asked, the Bakers decide to check the likelihood of their actually implementing this solution.

There is heavy *pressure* to make a decision, since Mr. Baker has accepted the new job and thus the existing system solution (i.e., apartment) is no longer viable. The *relative advantage* of the inner-city house over the existing system is thus very great, however; it is about equal in value with the condominium. With the Bakers being organizational players, and the concession to Mr. Baker's individual player desire to play tennis, there is now excellent *goal congruence* with the inner-city house solution. The *amount of behavioral change* is low with the house since the Bakers presently live in an inner city and the Bakers are classified as *skeptics* toward change.

The Bakers thus decide to go ahead with the *implementation plan* (step 9), by signing the contract, putting down a retainer fee, etc., for the inner-city house, setting up moving dates from their apartment in the Midwest, registering the children in the local Catholic school, buying a new couch for the living room, etc. Additionally, as part of the *acceptance test* for accepting the house, the Bakers have stated that (1) the house must be available by the 15th of August, (2) the crack in the backyard retaining wall must be fixed, etc.

In moving into this house in Los Angeles, the Bakers realize that their system will undergo much change. Things will be different from their life in the apartment in the Midwest. They decide they need to set some *systems goals* to monitor how their "new life" situation is going. From the evaluation matrix for alternative B (house), they use the expected performance as their goals. For example, on children's education, they expected above average performance. More specifically, the Bakers expect their children to get a solid religious foundation, plus good fundamental skills in "reading, writing, and arithmetic."

In setting these goals, the Bakers realize that certain tradeoffs are required with every alternative. For example, the crime rate in the inner city is high and certain precautions will have to be taken. Further, the aesthetics of the house is likely to remain relatively poor.

Once the goals are set and the Bakers move in, they monitor what actually takes place versus what they expected (problem-sensing, step 11). Assume six months from now that the children's education is worse than had been expected. Since there is a deviation between expected and actual performance, there are symptoms of a problem. If the Bakers decide the cause of the problem is the approach to education at the Catholic school, the Bakers now have a new subsystem problem. What is the best school for their children? The alternatives could be stay with the Catholic school, go to another private school, enroll the children in public schools, etc. In this case the Bakers would be keeping the same house, Mr. Baker is staying with the same job, etc.

Another situation might develop where after one year of living in Los Angeles, their son has become allergic to the relatively heavy smog conditions found in the inner city. Now the Bakers have a new systems problem which is, Where is the best place for the Baker family? Be sure to note that this isn't a rehash of the original problem. Because the systems context has changed, this is a different problem.

CONCLUSION

Through the housing example, we have taken a relatively simple situation to illustrate the systematic systems approach. The ten major processes have been highlighted in a step-by-step, integrative fashion. When this overall process is combined with the progressively comprehensive system study reports, the analyst has a systematic way to approach the solution to any systems problem.

Chapter Eighteen

Less Than Complete System Studies

Through the previous chapters of this text, the systematic systems approach has been developed and the rationale for each facet has been explained with examples given to emphasize various points. But we also have seen that systems problems can't really be "solved" in the traditional sense of that term. Further, we have noted the very heavy investment in time and cost to perform a solid system study. The major question that needs to be addressed in this chapter is how and when should less than complete system studies be performed?

SYSTEM STUDY RESOURCE/EFFECTIVENESS

We have stated that system studies should be performed in an iterative fashion which spirals toward an effective systems solution. This process is shown in Fig. 18-1.

As the study progresses, we see that there has been a move toward certainty (i.e., lessening of uncertainty) as shown on the right-hand side of the diagram. However, to accomplish this, an increasing commitment of cost and time is required.

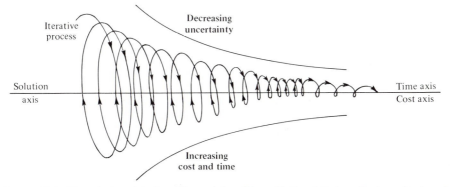

Figure 18-1 The spiral model of problem-solving. (From Shell and Stelzer, *Business Horizons*)

Since there is always a better solution to be found, when should one conclude a system study? In Fig. 18-2, the basic relationship between increasing system effectiveness and study resources is shown.

With increasing study resources of time, dollars, manpower, etc., the problem-solving group should be able to come up with increasingly effective system solutions. At any point the difference between the proposed solution and its expected system effectiveness, and the present level of effectiveness of the existing system, will be defined as the improvement.

In general, when the system study is initially performed, there is a fairly substantial investment in becoming acquainted with the system and establishing the systems context. This is depicted by area A in Fig. 18-2, which shows study resources committed, but little if any improvement in systems effectiveness.

Area B shows the next stage, where there is a fairly rapid improvement in system effectiveness with each additional commitment of study resources. The final stage is shown by area C, which shows a leveling off of system effectiveness with increased study resources of men, money, and time.

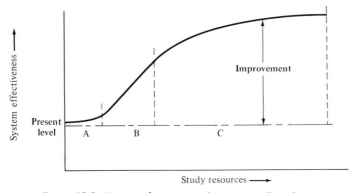

Figure 18-2 Impact of system study resource allocation.

However, the appropriate question is not will the system effectiveness improve with increased study resources, for this is generally true. But rather, will it improve enough to more than pay for the study itself? In Fig. 18-3, the general increase cost of a system study over time is shown as a dotted line and has been added to the previous graph of Fig. 18-2.

Up to point 1, the cost of the study more than exceeds the resulting improvement in system effectiveness. In other words, there is a start-up cost which must be borne. From point 1 to point 2, the system study is costing more and more but the resulting improvements in system effectiveness more than offset the cost. Beyond point 2, the system improvements are not enough to offset the increased cost of the system study.

Therefore, the effective CDM will insure there is at least enough time and resources to get past point 1. If there isn't, the system probably isn't worth studying. However, the CDM needs to stop the study prior to point 2. Where in the range between point 1 and point 2, the study should be completed depends on what other problems the analyst could otherwise be working on. The greatest improvement for the cost is found at point 3.

WAYS TO USE LESS STUDY RESOURCES

From Fig. 18-3, we have seen that the problem-solving group always needs to be aware of the relationship between possible increase of system effectiveness and the cost in resources for the study of the system. Ideally, they should try to stay between points 1 and 2 and preferably around point 3 in Fig. 18-3.

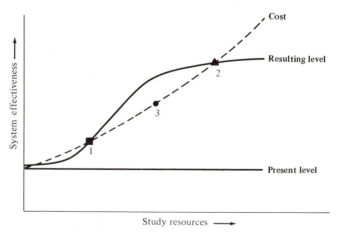

Figure 18-3 System study effectiveness/cost tradeoff.

We have seen in using the complete systematic systems approach that a great expenditure of resources in terms of time, manpower, etc., is required. Therefore, one conclusion would be to limit system studies to major problems of significant importance to a system. While this is possible, a more pragmatic approach is to tailor the structured systems approach to the problem at hand so as to approximate a point 3 situation.

For various reasons we may not want to do a full systems study. Problems which aren't that significant, or very significant but we have little hope of decreasing the uncertainty enough to make major improvement in the system effectiveness, or very significant problems but the time to decision is too short, etc., may lead us to consider the ways in which we can modify the systematic systems approach.

That is, how can we reduce the systems approach to a less time-consuming process, yet retain much of its problem-solving power? We will discuss ten ways the systematic systems approach can be simplified.

Minimize Writing

One of the most time-consuming elements in the systematic systems approach is the Final Report. This formalized, written document is the complete wrapup of the system study; the format of the report was shown in Table 10-4. While clearly there are many advantages to having this report, it is very time-consuming and therefore costly. Where practical, the analyst could report the final results in an oral presentation in which the various charts are displayed. That is, the evaluation matrix, systems utility function, systems simulation charts, and the preference chart would all be worked out in detail, but there wouldn't be any written discussion. All the previous reports—i.e., preliminary, feasibility, and evaluation—would be kept as is.

Simplified Evaluation Matrix

We have given much justification to show why the evaluation matrix (Fig. 12-9) needs to consider the feasible alternatives, weighted criteria, etc. When necessary to save time, however, it is possible to simplify the evaluation matrix by not explicitly considering the confidence levels. This in turn, would eliminate the discounted value and overall confidence items. What would be retained is shown in Fig. 18-4.

A further simplification would be to extract the dollar cost criteria (i.e., capital investment) and combine all the other criteria to be considered as showing primarily benefits. Then for example, when making a decision, the CDM could inquire whether it was worth say

Feasible Alternatives

	R	U	R	U	R	U	R	U
Criteria								
Total Value								

Figure 18-4

4,000 extra dollars to attain a 30 percent improvement in net benefits. See Fig. 18-5 and the matrix in class exercise 18-1.

Combine Simulation and Utility

Two of the most time-consuming aspects of the systematic systems approach are the development of the systems simulation charts and the systems utility function. Each has its critical bearing on the re-

Feasible Alternatives

	R	U	R	U	R	U	R	U
Criteria (Noneconomic)								
Total Value								

Criteria (Economic)								

Figure 18-5

sulting system study effectiveness. However, much time can be saved while retaining some of the effectiveness by combining these two aspects.

In the evaluation matrix, each alternative can be rated on every criterion by using terms like excellent, average, barely acceptable. However, no explicit rationale would be given for the underlying value set nor the supporting evidence for how the expected performance was derived.

Later, each term can be given a point value. For example, excellent is 10 points, average is 5 points, etc. Sometimes in the overall ratings symbols like ++, +, 0, −, −− are used where 0 would indicate normal performance, and ++ is performance in the top 10%, for example.

However, if the first alternative is rated on all criteria, then the next alternative is rated, etc., there would not be any real comparison base. To ensure less bias, all alternatives should be rated on one criterion at a time. Then the overall value is calculated.

Limit Scope of Analysis

As we have studied the system and developed the systems context diagram, we are eventually forced to make a decision on the scope of the study. We definitely don't want to arbitrarily narrow the focus of the study down to a well-defined, but essentially useless problem statement. At best, this would provide a solution with no impact on overall system effectiveness and at worst, could mislead us into developing a solution which is worse than the present situation.

So in essence, we want to take on as "big a bite" as we can handle with the allocated study resources. If this is less than what is needed, then we can still tackle the same problem, but hold more of the systems context variables constant. This, in turn, reduces the study complexity since it reduces the number of variables and limits the interface considerations.

The planning horizon could be reduced. For example, in the case of marriage, which "should" have a 50-year planning horizon, we could consider 5 years. The number of alternatives to be considered could be reduced, more constraints could be added, etc.

No Explicit Systems Utility Function

As we have noted, one of the most time-consuming aspects of a systems study is the development of an explicit systems utility function. To have its greatest impact, the systems utility function should be developed from a comparison base which is meaningful. This, in turn, requires the conceptual study of a minimum of 30 feasible alternatives and preferably 100 alternatives.

This is not always practical, nor many times even possible, so the analyst could generalize only from those alternatives under study. While this clearly will bias the results, it does save considerable time.

Perhaps a better approach would be to establish the average value for each performance measure from a large set of alternatives. If time is available, then the extremes could be generated—that is, the exceptional and barely acceptable performance levels. Where possible, the crucial point of the system utility function should be retained, which is an explicit yardstick showing the relative value of performance inherent in a certain situation.

No Preference Chart

While the preference chart doesn't really take a great deal of time to prepare if there is only one CDM, it can become involved if we are developing a consensus among multiple decision-makers. Instead of doing a one-by-one comparison, time could be saved by just asking people to rank in order the criteria. From this, a rough grouping of criteria by importance could be made.

To get numeric weightings, the analyst could ask the CDM(s) to assign weights which add to 100. This tends to force a tradeoff and gives reasonably good relative importance indications. In no case should the analyst assume every criterion is equally important. While this assumption simplifies the analysis, it misses the whole point.

Limit to Major Criteria

We have stressed the need for multifaceted evaluation of system solutions in the study of complex situations inherent with cybernetic systems. Therefore, the analyst shouldn't reduce the number of criteria arbitrarily to save study time.

If there has to be a reduction in the number of criteria on which the alternatives are evaluated, the less important criteria should be dropped first. The minimum level of acceptance is to limit the detail study to those criteria which are classified as very important. The important and moderately important criteria can be considered implicitly when the decision needs to be made on which feasible alternative is best overall.

In no case do we ever want to limit the evaluation to one criterion or dimension. There always have to be at least two (benefits and costs).

Limit Alternatives Considered

The wider the range of alternatives considered and the greater the number, the better the chance of selecting a "good" solution. In the systematic systems approach, we suggested that adaptavising was more practical than searching for optimization. But when time is limited, the number of alternatives to be evaluated in depth can be further reduced.

At the beginning of the study, however, a wide range of alternatives should be considered, with very tight constraints then applied to quickly get down to the "good" solutions. As a minimum, two alternatives should always be considered where possible: the proposed solution and the existing system design. The existing system solution can be used as a comparison base. In cases where the existing system is infeasible and can't be reasonably modified to make it feasible, the analyst can use the proposed solution as "the" alternative in a satisficing mode. For comparison the proposed solution can be varied in cost for example by +10% or −10%.

Not Marketing System Solution

Most system studies are completed without the analyst ever realizing that "marketing" or promoting the recommended solution is a valid aspect of the systems approach. This, in turn, is a major reason why most study recommendations are not implemented. However, there are a great number of situations where the analyst doesn't have to be involved in this very demanding and time-consuming phase of systems problem-solving. As was pointed out in chapter 15, when the CDM has the relatively greater expertise the analyst just needs to "report the facts." Further, in situations where the main decision-makers are organizational players or there is inherent goal congruence, the time spent in marketing can be limited.

In the remaining situations, the analyst needs to market the study conclusions. If time isn't available for this, the probabilities of the solution acceptance decrease significantly. But this may be the price that has to be paid, when there are heavy demands on the analysts and CDM to handle multiple projects.

Limited Talks with CDM

The last alternative way to do a system study with less than adequate resources of time, manpower, etc., is to limit the discussion with the CDM. The iterative, spiraling problem-solving philosophy is a major part of the systematic systems approach. The preliminary, feasibility, and evaluation reports reflected this method and with them is the explicit need to have periodic discussions with the CDM or problem-solving group. While these reports and discussions have been designed to be relatively short in the amount of time required, they do take time.

In extreme cases, this may be time that just isn't available or possibly the CDM doesn't want to talk to the analyst or maybe he subscribes to the completed staff work concept of chapter 10. By doing the study on his own, the analyst will probably save much time, but the value of the conclusions will most likely be much less than with the iterative approach.

PREFERRED ORDER OF COLLAPSING STUDY

We have discussed ten ways of reducing the resources necessary to perform a less than complete system study. The potential amount of time that could be saved was pointed out, but the relative effect of removing or reducing various steps of the systems approach needs to be evaluated.

If time were the most limited factor (see Table 18-1), the analyst should try to reduce the formalized writing (factor A) and have no systems utility function (factor E). This would imply a system study without an explicit systems utility function and the results would be communicated to the CDM verbally with the evaluation matrix, systems simulation charts, and preference chart included. The less effective time-savers are the removing of the preference chart (factor F) and the combining of the systems simulation chart and systems utility function (factor C).

Table 18-1   Time and value considerations.

Factors	Time Saver	Value of Decision	Time/Value Rank
A. Minimize writing	Significant	Reduced	1
B. Simplified evaluation matrix	Above average	Reduced	2
C. Combined simulation and utility	Less than average	Slightly reduced	4
D. Limit scope of analysis	Average	Reduced	5
E. No systems utility function	Significant	Greatly reduced	3
F. No preference chart	Much less than average	Slightly reduced	8
G. Major criteria	Greater than average	Greatly reduced	6
H. Limited alternatives	Greater than average	Greatly reduced	7
I. No marketing solution	Greater than average	Significantly reduced	9
J. Limited talks with CDM	Average	Significantly reduced	10

But time saved isn't the only consideration; rather the analyst should consider the potential impact on the worth of a decision derived without explicit use of any of the ten factors. In this case, the least impact is felt when factor C, combined simulation and utility, is carried out. Simplifying the evaluation matrix (factor B) or reducing the formalized written report (factor A) causes a reduced value of decision. The most impact is generated when the study results aren't marketed (factor I) or when the analyst has limited talks with the CDM (factor J).

Time/Value Rank

When the analyst can't do a complete system study, he should consider both the time to be saved by limiting certain considerations and the potential impact on the resulting worth of the decision.

The right hand column of Table 18-1 shows the preferred order of collapsing a study, using a time-saved to value-of-decision ratio. This is recapped in Table 18-2. When time is limited, the analyst should forgo, where possible, the formalized writing and go to a simplified evaluation matrix. The last things to give up are the marketing of the solution and the iterative talks with the CDM.

SUMMARY

From the discussion throughout the text, we have developed a complete methodology called the systematic systems approach which guides us in solving complex systems problems in a systematic way.

Table 18-2 Preferred order of collapsing study.

Minimize writing	1st
Simplified evaluation matrix	2nd
No systems utility function	3rd
Combine simulation and utility	4th
Limit scope of analysis	5th
Major criteria	6th
Limited alternatives	7th
No preference chart	8th
No marketing of solution	9th
Limited talks with CDM	10th

This chapter has shown how to keep the overall process intact, while tailoring the depth of what is done to the resources available for studying any particular systems problem.

To further see the power and usefulness of the systematic systems approach, we need to see how it compares to other contemporary approaches or techniques for problem-solving. We will show this in the following class exercises. After you have completed these exercises, see if you agree with the statement that most other problem-solving methods are subsets of the systematic systems approach.

Class Exercise 18-1

MODIFIED EVALUATION MATRIX

Assume you are a computer consultant who has been asked to aid a small business in the selection of a minicomputer system. You are working with a manager who has little background using computers. Further, he is just interested in the bottom line, and uses a very decisive management style.

Because of these considerations, you have decided that the evaluation matrix will be the primary document with which to brief the manager. There is no other written report, because the manager isn't willing to pay for you to write it. In the interest of time you have not made explicit the preference chart, system simulation, or utility function. Because the manager doesn't like to hear about uncertainty on the part of the consultant, you have deleted the confidence levels. As a result, what is shown in Fig. 18-6, is a simplified evaluation matrix.

		Fiscal		Medical		Dynamic	
		R	W	R	W	R	W
3.5	Applications software	5	17.5	6	21	6	21
	Risk	4	14	6	21	6	21
2	Price	2	4	1	2	10	20
	Ease of use	9	18	8	16	7	14
1	Growth capability	6	6	8	8	8	8
	Service	5	5	8	8	8	8
	Computer system	2	2	7	7	6	6
	Total score		66.5		83		98
	Total score w/o price		62		81		78
	Price		$80K		$84K		$60K

Figure 18-6 Simplified evaluation matrix.

On total score the Dynamic computer system looks best with 98 points. However, since the CDM was specifically interested in the bottom line, price was taken out of the criteria and total score figured on the remaining six criteria, without price.

This yielded a different picture, with Medical Systems getting 81 points of net benefit and Dynamic 78 points. These potential benefit scores were then put over the system price.

1. How could the criteria be ranked and weighted without a preference chart?

2. Explain what a rating of 10, 7, and 2 means without an explicit systems utility function or systems simulation chart.

3. What is the implication of an analysis made on total score without the explicit knowledge of the confidence levels? Why wouldn't a manager want to know about this uncertainty?

4. Is it reasonable to extract price out as a special consideration? In this case, is it worth 24,000 additional dollars over a five-year period to receive 3 more points (81 versus 78) of benefit?

Class Exercise 18-2

REASONED ANALYSIS

The following example is typical of a reasoned analysis that people might use in solving a personal problem. In this case, the advertisement is trying to persuade you to go with XYZ as the best cigarette brand for you to smoke.

Why Do You Smoke?

With what you've been hearing about smoking these days, you probably wonder sometimes why you smoke at all. Yet you enjoy it. Because smoking a cigarette can be one of those rare and pleasurable private moments. And the chances are you don't want to give up any of that.

Which brings us to the XYZ brand. XYZ is the cigarette for people who have come to realize that most cigarettes that give them the flavor they want also give them a lot of the 'tar' and the nicotine that they may not want.

XYZ is the cigarette for people who've found that most low 'tar' cigarettes don't give them anything at all. The thing that makes XYZ special is that its filter is based on a new design concept that gives smokers the flavor of a full-flavor cigarette without anywhere near the 'tar' and nicotine.

Now we don't want to suggest that XYZ is the lowest 'tar' and nicotine cigarette you'll find. It isn't. But it sure is the lowest one that will give you enjoyment.

And that's why you smoke. Right?

Questions

1. What is the problem? Who is the CDM?
2. What are the criteria suggested that should be used? Should there be others?
3. What are the alternatives? What is the best solution?
4. How could this narrative analysis be looked at as a very simplified form of the systematic systems approach? What has been collapsed? Are there any inconsistencies?

Class Exercise 18-3

NET PRESENT VALUE

There are many types of simple formulas that are used to evaluate the worth of various alternatives. One of the key concepts in the literature of finance is capital budgeting. It is used to assess the present value to an organization of alternative long-term investment proposals, as exemplified below.

Which Project Is Best?

Assume there are three investment proposals. Investment 1 requires a cash investment now of $18,000 in fiscal year 1980 and it will return net cash flows of $5600 dollars for each of the next five years (i.e., 1981, 1982, 1983, 1984, and 1985). Investment proposals 2 and 3 are as shown below.

Investment Proposals

Year	#1	#2	#3
1980	−$18,000	−$18,797	−$16,446
1981	+$ 5,600	$ 4,000	$ 7,000
1982	+$ 5,600	$ 5,000	$ 6,000
1983	+$ 5,600	$ 6,000	$ 5,000
1984	+$ 5,600	$ 7,000	$ 4,000
1985	+$ 5,600	$ 8,000	$ 3,000

As an aid to decide which is the best investment, the field of finance uses what is known as the cash flow discount (CFD) method. The essence of the CFD method is that one is not indifferent between receiving $1,000 today or getting the $1,000 five years from now. The money received today could be invested to earn a return. Therefore, to get a common comparison figure among investments, all cash returns are discounted back to their cash worth in the starting year.

If money could be invested at a net rate of 4%, (i.e., interest rate minus inflation rate), the present value of each alternative is shown below.

	#1	#2	#3
Present Value	$6,930	$7,565	$6,162

If the alternatives are ranked by their present value, that investment which has the greatest present value is considered the best. Therefore, investment #2 is the best solution.

Questions

1. What is the problem? What are the alternatives? Who is the CDM and for what system?

2. What is the planning horizon? What is the criterion used? What assumptions are made about the system benefits of the various project proposals? What is the implicit assumption made on the accuracy of the projected cash flows?

3. What is the risk of losing money on any of these projects? Should it make any difference that project 3 could give employment to 100 workers that were recently laid off? Does it make any difference that the boss (CDM) considers project 1 to be his pet project? In what way is investment #3 enhanced by the fact that it contains money for installing needed anti-pollution devices?

4. Discuss the net present value method as a simplified part of the systematic systems approach.

Class Exercise 18-4

LINEAR PROGRAMMING

One of the most widely used methodologies of the fields of operations research, management science, and decision sciences is the powerful tool of linear programming. Look up an example given in any of the current textbooks used in those fields. Then answer the following questions.

1. Identify the problem to be solved. Who is the CDM? What is the planning horizon?

2. What are the alternative solutions? How is it determined which is best? Does linear programming use the method of optimization, satisficing, or adaptavising?

3. What relationship does the objective function have to the evaluation matrix? What implicit assumptions are made about the information used? How are nonquantitative criteria handled?

4. Once the best solution is determined, what procedures are given for communicating the results to the CDM? What form of systems planning is used to ensure implementation?

The Hospital Billing Case *

The purpose of this five-part case is to give the student further practice in applying all the basic aspects of the systematic systems approach. No previous knowledge of computers, patient billing systems, or hospital operations is required to understand the essential aspects of the case.

While the information and context surrounding the case have been developed to simulate a realistic problem-solving situation, much of the detail has been left out. Don't get bogged down in this regard; keep an overall perspective. Remember, the purpose is to see all the steps.

PART I

Harry Schmaltz is the administrator of a small hospital located in a suburb of Topeka, Kansas. The hospital uses a computer exclusively for processing their patient's bills. Each month about 12,000 billings are processed on the hospital's Control Data Corporation computer. Although the present billing

*This case is a modified version of the one developed by Steve Zalewski as a senior project at California State Polytechnical University, Pomona.

All information provided in the case is fictitious. The company and computer manufacturer names are only used for the purposes of student analysis and discussion.

system seems to be working adequately, the lease on the hospital's computer is about to run out and Harry wants to check out some other possibilities in the hopes of reducing costs or increasing efficiency of the billing system.

When a patient goes to the hospital for treatment, a bill is made out by a member of the hospital staff. This bill contains information such as patient name, patient number, type of treatment received, patient balance, etc. The information from the patient's bill is then keypunched onto cards and processed through the computer.

The present billing system called Ajax was previously designed by Jerry Jones, the hospital's present DP manager, and is maintained by two programmers. During the last 5-year computer lease, Mary Smith, the hospital's controller, has constantly monitored the costs for processing the patient's bills. By considering such things as the lease cost of the computer, DP center wages, overhead, etc., Mary has determined that each patient's bill that is processed through the computer costs 51¢. Ninety percent of the bills that are processed are paid by Blue Cross, Blue Shield, or some other medical insurance company.

Harry assigned the task of researching the various types of computers and billing methods that the hospital could use to Jerry, with the understanding that $80,000 was the most money that could be made available for patient billing next year. Because of growing competition from other hospitals and the fact that this hospital serves a relatively stable community, Harry estimates that the patient load over the next five years will remain about the same.

Jerry began contacting various companies that sold or leased computers and other hospitals that used computers for their billing. After one week of talking to different companies, Jerry found that in addition to the Ajax billing system which the hospital presently uses, three other possibilities would work in the hospital's billing procedure.

The first of these was a billing system leased by the Medical Billing Company. The software package was designed to run on a medium-size minicomputer manufactured by National Cash Register. Jerry figured that in addition to the yearly lease cost of the NCR computer, a $14,000 conversion cost would be incurred in the first year only. This conversion cost is for employee familiarization with the new system, new billing forms, physical modifications of the computer room, etc.

Computershare, Inc., has a billing system which uses an IBM computer with medical billing programs available. Information that would normally be handwritten on the bill would be entered instead on a computer terminal. A terminal is a device that would be installed at the hospital and connected to a computer located in Topeka. The terminal would enable the hospital to request any information on file about any patient treated by the hospital. Computershare charges 43¢ for the processing of each patient's bill. A one-time cost of $10,000 would be incurred by the hospital in order to convert to the Computershare billing system.

Universal Processing, Inc., uses a large Burroughs computer to process their medical billings. Each day at 9:00 A.M. couriers leave the downtown Topeka office of Universal and pick up the patient's bills at hospitals in their designated areas. These bills are returned to Universal for processing. After being processed, updated patient's accounts are returned to the hospital in 24 to 36 hours. Universal will charge the hospital 48¢ for the processing of each bill. Jerry figured it would cost about $7,500 to convert to this system.

Questions

From the information about the case, answer the following questions:

1. Determine what the problem is and establish what is its cause(s). What type of problem is this?
2. Describe the system context.
3. What is the composition of the problem-solving group? Who are decision-makers and who are analysts? Who is the CDM?
4. What is the scope of the study? More specifically, what is the planning horizon? the time to decision? and the amount of study resources available?
5. What constraints or requirements are involved with each potential alternative?
6. Describe each alternative that is being considered. What must any systems design do to be considered a feasible alternative?
7. What are the most appropriate system objectives that apply to this situation?

PART II

Jerry made up the following chart to summarize what he had learned so far:

Mary's calculation of the processing cost of Ajax,
the present billing system was 51¢/bill. . . . . . . $73,440/yr.

Medical Billing @ 45¢/bill . . . . $64,800/yr.
     Conversion Cost $14,000

Computershare, Inc., @ 43¢/bill . . . . $61,920/yr.
     Conversion Cost $10,000

Universal Processing, Inc., @ 48¢/bill. . . . . $69,120/yr.
     Conversion Cost . . . . $ 7,500

When Mary looked at the chart she noticed that Ajax was the most expensive of the four choices for processing each medical bill. However, since no conversion costs were encountered in the Ajax system, she thought it would be a good idea to figure out how long it would take to recover the conversion costs based on the cheaper processing methods. To do this, she subtracted the yearly processing cost of each choice from the yearly processing cost of the Ajax system. She then divided the conversion cost by that remainder to give her the time to recover the conversion cost. Mary then determined that the time to recover conversion costs was 0 years for Ajax, 1.6 years for Medical Billing, 0.9 years for Computershare, and 1.7 years for Universal Processing.

Of interest to the group was how long it took to establish each patient's new account balance. With the present Ajax system, a combination of manual and computer processing needed to take place. Using this as the norm, Jerry, the DP manager, felt that with Computershare each patient's balance could be determined almost immediately by requesting the information on the hospital's computer terminal. The group agreed that this was exceptional compared to what other systems being considered offered.

Universal Processing gave the same information but since they used a courier service, it took 24 to 36 hours for that information to be received by the hospital. The group conceded that this was more time than what other systems normally offered.

With the Medical Billing system the account balance could be attained very quickly. The only real delay was that the computer couldn't do multiple jobs at the same time. Although the computer would be processing other jobs much of the time, the group felt the overall response would be above average.

To summarize the data obtained, Jerry prepared the chart shown as Fig. A-1.

	Yearly Operating Cost	Conversion Cost	Time to Recover Conversion Cost	Time to Receive Updated Patient Balances
Ajax	$73,440	$0	0 years	Average
Medical Billing	$64,800	$14,000	1.6 years	Above average
Computer- share	$67,920	$10,000	0.9 years	Exceptional
Universal	$69,120	$ 7,500	1.7 years	Below average

Figure A-1

Questions

8. Mary recommended Computershare, Inc., as the best choice for a hospital billing system. She said, "What's the problem? Since all choices will work, the best one is the least expensive." Do you agree or disagree? Why?

9. Is there any other information that could be as important as cost in this decision?

10. If the Medical Billing Company was selected as the best alternative, what specific steps would have to be taken to implement this problem solution? How does this compare to the implementation procedure required if Ajax was selected as best?

11. What would the systems goals become for the Medical Billing system after the desired solution was implemented?

12. What billing system is best if one considered only:

   (a) yearly operating cost?
   (b) time to receive updated patient balances?

13. Why is it desirable to have a low time to recover conversion cost? What is this approach usually called in financial theory?

14. If you had to make a decision based on the information given so far in the case, which alternative do you feel is best? Justify your answer.

PART III

Harry opened the second meeting with the statement, "How sure are we that the information in this last chart is accurate? I know that our yearly costs are more accurate than the conversion costs for these choices, since we know about how many bills we process. But the conversion costs seem to be less certain than yearly costs because some of that conversion cost relates to how long it will take our staff to learn how to use the new billing system. And how confident are we that an individual can learn the system in 5 hours or 10 or 20?"

Although they hadn't thought of it that way, Mary and Jerry felt that this was a valid point. After discussing this idea, the group decided to establish some measure of how confident the group was of a particular piece of information. They decided on four confidence ratings: (1) *very confident*—this rating would be used for information that was essentially fact; (2) *confident*—this would be used when there was greater than a 50/50 chance that the information was true; (3) *low confidence*—this rating indicated the piece of information was something more than an "educated guess"; (4) *no confidence*—this rating would be used when little or no substantiation could be given to the information. Basically it was just a guess.

After agreeing on the confidence ratings, Jerry expressed reservations about having the accounts processed outside the company. He knew of many instances where patients were being billed who had no treatment because "outsiders" were able to get the names and account numbers of those patients and use them to get free medical treatment for themselves. Consequently, the hospital lost all revenues from the treatments. In addition, in order to balance the accounts properly, an extensive audit had to be undertaken. This could cost the hospital from 10 to 20 percent of their yearly DP budget. An unnecessary audit costing $7,000–$15,000 in this case, was to be avoided if at all possible. Another difficulty of inadequate security was the loss of revenues due to the patient's dissatisfaction with the billing department of the hospital, causing patients to seek medical treatment elsewhere.

Jerry also related how hospital operations could be catastrophically affected if daily patient bills were somehow "misplaced." The reconstruction of those bills was not only difficult but sometimes impossible. Those at the meeting agreed that the chances of these things happening must be minimized if the hospital was to remain financially solvent and respected in the community.

With both the Ajax and Medical Billing systems, the processing would take place at the hospital itself. This would result in better security because all employees who came in contact with this vital information could be thoroughly checked and evaluated by the hospital's personnel staff. Consequently, they were very confident that those two systems offered the best security.

Harry, the hospital administrator, had worked for a hospital in the East for five years which used Computershare, Inc. billing. He was confident that Computershare did an above-average job of ensuring the security of their users' files, but the group agreed that the best security was that assured by their own staff.

When Jerry finished talking to the marketing representative of Universal, he felt that Universal did an above-average job of security. Harry said he had a nephew who used to be a courier for Universal and Harry decided to give him a call. After talking to him, Harry found out that Universal used a private courier service that paid little attention to the quality of personnel that it hired. His nephew said that no security checks were made on him and he knew of no one who was required to go through any. Consequently, the group had little confidence in the assurances of the marketing representative.

Jerry then brought up the possibility of the effects of computer down-time on operations at the hospital. "In an installation of this size," he said, "it can be expected that the computer would go down an average of twice a week." And Jerry was dissatisfied with the amount of time it took for a service technician to arrive to repair the computer when it was down. On their present CDC computer, it took an average of five hours for a service technician to arrive. During this time, obviously, all processing stopped and operations were disrupted. Jerry felt that this amount of time lost was barely acceptable and the group agreed.

With the National Cash Register maintenance service, a service technician was guaranteed at the user's site within three hours. By speaking to several users of other makes of computers, the group was confident that three hours was a normal time to wait for a service technician.

Universal had perhaps the best provisions for computer down time. They had been providing their medical billing service to hospitals for many years and were aware of the problems that downtime could cause in the hospitals they served. They had an innovative staff that prided themselves on their consistency of computer operations. They achieved this by using a large-size Burroughs computer to run the medical billing programs. In the event of a malfunction, however, the processing could be taken immediately by another Burroughs computer. In fact, the company guaranteed that computer downtime would cause no delay in the normal turnaround time for delivering processed data to the client. This would be explicitly written in the contract.

Since Computershare used a large-size IBM computer, the group was confident that a repair technician would be on the site in two hours or less because, as all were aware, service had made IBM one of the best known in the field for many years. So there would probably be minimal downtime problems for the hospital with the Computershare system.

Harry wanted to know how confident the other members were on the yearly operating cost and conversion costs. They agreed that because of inflation and the increasing cost of labor and materials, they were not very confident that these costs would remain the same as in the past. But since their present Ajax billing system had been in use for over five years, they felt more confident of its operating and conversion costs than of those same costs for the other three alternatives.

Following the meeting, Jerry went home to prepare the chart shown as Fig. A-2.

Questions

15. Which criteria are qualitative and which are quantitative? Give an interpretation for what the particular confidence level means on the criterion of "time to receive updated patient balance." What alternative overall has the greatest confidence ratings? Why is this?

16. (a) Mary, the controller who was a C.P.A. with an MBA in finance, protested the information Jerry had presented. She stated, "This information is mostly superstition, since only dollar costs have any affect on the accounting statements."

    (b) "Furthermore, what's all this confidence bit? Dollars are dollars and there is no need for this qualifying information. Either the numbers are correct or they are not!" State why you agree or disagree with her views.

17. Harry suggested that the yearly operating cost and the initial conversion cost for each alternative be compared using present value theory. Then the rest of the criteria can be considered as intangibles. Using this method, decide which alternative is best.

18. What information isn't being used in the method suggested by question 17? Comment on the statement, "To try and use all

	Yearly Operating Cost	Confidence	Conversion Cost	Confidence	Time to Recover Conversion Cost	Confidence	Time to Receive Updated Patient Balances	Confidence	Security	Confidence	Delay To Wait For Service	Confidence
Ajax	$73,440	C	$ 0	VC	0 yrs	VC	average	VC	outst	VC	5 hrs	VC
Medical Billing	$64,800	LC	$14,000	LC	1.6 yrs	LC	average	LC	outst	VC	3 hrs	C
Computershare	$67,920	LC	$10,000	LC	0.9 yrs	LC	outst	LC	abv avg	C	2 hrs	C
Universal	$69,120	LC	$ 7,500	LC	1.7 yrs	LC	bel avg	LC	abv avg	NC	0 hrs	C

Figure A-2 Initial Summary Chart.

the qualitative and quantitative criteria along with the accuracy indications seems like comparing apples and oranges."

PART IV

When Jerry brought his chart to the next meeting, he expected the matter to be discussed and a decision made based on this chart. But Harry brought up the earlier discussion about the greater importance of security. "How do you feel about this business of data security, Jerry?" "I think it's critical. It's certainly more important than the time it takes to recover our conversion costs. In fact, I think it's as important if not more so than any of the other criteria." Mary and Harry both agreed. "But does our chart reflect that importance?" Jerry said. The chart treated all of the criteria the same.

For a moment, the group was at a loss as to how to handle the criteria of security. "How much more important is it?" Mary queried. "I don't know, I guess maybe three times as important," Jerry retorted. Harry concurred. Then it hit Jerry. "I'll make up another chart which will give numerical values from barely acceptable to exceptional performance, and then those criteria that are more important can be multiplied by some factor that reflects their relative importance." In the case of security, the factor will be 3, because security is three times as important as other factors.

Jerry further pointed out that his billing programs would only work on Control Data computers, without completely rewriting them. This, he explained, was due to the fact that the programs were written in a computer language which was unique to that computer manufacturer. If, at a later time, the hospital decided to get a computer from a different manufacturer, it would take 800-1,000 man-hours to rewrite the present programs to run on the new computer. In a sense, the hospital was "locked-in" to using Control Data computers. This fact made them less flexible to take full advantage of their options in the future.

Universal Processing has been providing their medical billing service to hospitals for over ten years. Since they are primarily a service organization, they cater to the needs of their users. Jerry was assured by the representative of Universal that any legislative changes during the period of an existing contract would be Universal's responsibility and would result in no additional cost to the user during the life of the contract. From talking to various Universal users, the group felt that with Universal, the hospital's flexibility of adapting to changes in the external environment would be maximized.

Medical Billing's software is written in a programming language common in business (COBOL). The language, with minor modifications, can be run on almost any of the latest computers. If the hospital were to choose this system, their number of future choices of computers would be increased substantially over their present system. Jerry estimated the amount of time to convert the Medical Billing system from one computer to another at from 60-80 man-hours.

Mary then mentioned that she had read in the morning paper that Congress, under pressure from an insurance lobby, passed a bill requiring that in five years all medical payment claims would have to be in a form readable by optical-character-recognition machines. This requirement implied that a specific OCR font had to be used and very tight printer specifications had to be met. The hospital group wanted to take advantage of these OCR requirements well before the deadline.

At a local bar, Jerry talked informally with a sales representative of Computershare about these new OCR requirements. The rep stated that his company didn't want to spend any money modifying their medical billing programs or putting in special printers before it was necessary in five years. He felt this was so because the company considered hospitals to be a low-priority account. The hospital group decided to completely eliminate Computershare from further consideration, after hearing this.

Mary posed the question, "What if we had signed with Computershare last week?" They all agreed that a substantial loss of control over their operations would have resulted. After they fully considered how much of an effect this outside factor could have had on the functioning of the hospital, they decided that the flexibility to adapt to changes in the environment was just as important as security so they decided that the "multiplier," like that for security, would be 3 also.

It was now becoming apparent to Jerry that they needed a way to summarize what the problem-solving group felt was the desired performance of a medical billing system. This would include which facets surrounding a billing system were most important from an overall hospital perspective. Additionally, what was considered to be barely acceptable to exceptional performance on each of these criteria needed to be specified.

This summarization chart, which Jerry coined the "systems utility function" is shown as Fig. A-3.

Questions

19. Harry liked this chart, but he wanted to be sure he knew how to interpret it. So he asked the following question: "What is average machine specificity and how did you determine it:" What would your answer be?

20. Mary was more critical of this procedure. She asked the following very penetrating question: "Jerry, you say the systems utility function represents the desired performance of a medical billing system. Is that for all hospitals or just ours?

Rating	0	1	2	3	4	5	6	7	8	9	10
	Barely Acceptable		Below Average		Average		Above Average		Exceptional		

Flexibility

1. Minimum length of lease

Rating	Barely Acceptable	Below Average	Average	Above Average	Exceptional
Minimum length of lease	10 years	5 years	3 years	2 years	1 year
2. Machine specificity	Programs must be run on a specific model of a specific computer	Programs must be run on a specific computer	Programs must have 100-500 man-hrs conversion time to run on a different computer	Programs must have less than 100 man-hrs conversion time to run on a different computer	Programs able to run on any computer
Security	No control over those who handle data	Less than $\frac{1}{2}$ of the people who have access to data are subject to security checks	More than $\frac{1}{2}$ of the people who have access to data are subject to security checks	All people who have access to data are subject to some form of security check	Extensive security check of all people who handle data at user's location

Delay to wait for service	4-5 hours	3-4 hours	2-3 hours	1-2 hours	No delay
Operating cost per year	More than $71,000	$67,000-$71,000	$64,000-$67,000	$60,000-$64,000	Less than $60,000
Conversion cost	More than $13,000	$11,000-$13,000	$ 9,000-$11,000	$ 7,000-$ 9,000	Less than $ 7,000
Years to recover conversion cost	3.0	2.0	1.0	.5	0
Time to receive updated patient balance	More than 2 days	1-2 days	6 hours-1 day	1-6 hours	Less than 1 hour

Figure A-3 Systems utility function.

Further, isn't our consensus, as the problem-solving group, most likely biased?" Determine your answers to these points.

21. Mary asked, "Why couldn't all dollar costs (operating cost per year and conversion cost) be combined and considered as one criteria?" Your reply is?

PART V

Harry now saw what Jerry was getting at. He said, "Since we can estimate the expected outcome for any particular medical billing system, and we have through the systems utility function what is considered as the desirable levels of performance for our hospital, all we need to do is just put the two together to establish which billing alternative is best." "Why don't you, Jerry, develop another chart to show this comparison explicitly? And we can call it an evaluation matrix."

Jerry worked up the chart shown as Fig A-4 which compares the overall disability of three billing systems.

Alternatives

		Ajax			Medical Bill			Medical Processing	
Criteria	R	U	C	R	U	C	R	U	C
3 Flexibility	5	15	C	6	18	C	8	24	VC
Security	9	27	VC	9	27	VC	6	18	LC
2 Yearly operating costs	1	2	C	6	12	C	3	6	C
Conversion costs	10	20	VC	1	2	LC	9	18	LC
Service delay	1	1	VC	4	4	C	10	10	C
1 Time to recover conversion cost	10	10	VC	4	4	LC	4	4	LC
Time to receive updated patient balances	5	5	VC	5	5	VC	3	3	C
Total Value		80			72			83	

Figure A-4 Evaluation matrix.

Questions

22. Mary, upon looking at the evaluation matrix, said, "Universal Processing must be the best alternative, since it has the highest total." Why do you agree or disagree?

23. Harry pointed out that no specific use was being made of the accuracy of information indicators. Jerry agreed and said, "Why don't we sort of discount the ratings by our level of confidence?" They all agreed this was a good idea.

    Determine what the overall alternative rankings would be if you used .9 for very confident, .6 for confident, .3 for low confidence and .1 for no confidence.

24. Using the discounted rankings, the group decided that Ajax, the present billing system, was best. (a) Do you agree? (b) What tradeoffs are involved with the Ajax systems?

25. Harry was furious. "Do you mean to tell me that after all these meetings, that we have decided to stay with our present setup? What a waste of time and effort." Tell why you agree or disagree with Harry's conclusion.

Just then a salesman from the hospital division of Honeywell Computer Systems walked in. He stated that he had heard the hospital was considering switching to a new medical billing system. Harry stated, "We have just spent the last three weeks looking into this problem, and we decided 15 minutes ago to stay with what we presently have." The salesman said this division of Honeywell was very competitive and aggressively going after new business. Jerry stated, "Since we already have the systems utility function completed, it would be relatively simple to compare the Honeywell system with our decision about Ajax." They all agreed.

Jerry then took the salesman to his office, where he proceeded to ask him specific questions.

Questions

26. What specific questions should be asked of the Honeywell salesman?

27. Assume that after these questions had been answered, Jerry did a quick calculation of the Honeywell alternative and came up with a discounted total score of 58. (a) What would you say to the salesman now? Jerry told him that if he could reduce the operating cost per year using the Honeywell system from $65,500 to less than $60,000, he could have the contract. (b) Discuss this reasoning. (c) If the salesman stated

there was no way he could do that without losing money, what would you say to him?

28. Assume it was decided to stay with the Ajax system. What implementation plans are necessary? What will the systems goals be for this system over the planning horizon? What happens if they aren't met?

Glossary

Above average performance is the level of performance indicated by the 70 to 90 percentile. This performance is beyond what could be normally expected from systems designs in this category.

Abstraction is the degree to which something differs in representation from the actual situation.

Adaptability of goals reflects whether the goals of the system are fixed or can be changed depending on the environmental condition and state of system learning.

Adaptavising is a method of problem-solving in which one continues to check viable (feasible) alternatives until the perceived costs of further search equal or exceed the poten-

tial benefits gained from a better system solution.

Alternatives are the different courses of action or approaches which could be used to solve a particular systems problem.

Alternative rankings conclude for the set of system designs to be evaluated, what their relative rank is according to overall worth.

Appropriate systems level determines the system boundaries as a function of the composition and authority of the problem-solving group.

Average performance is the level of performance indicated by the 30 to 70 percentile. This performance is what could be normally expected from systems designs in this category.

349

Bad performance describes those outcomes which give less than the normal or usual performance in a particular situation.

Barely acceptable performance is the level of performance indicated by the lowest 10% or 0 to 10 percentile. This performance is well below what could normally be expected.

Below average performance is the level of performance indicated by the 10 to 30 percentile. This performance is below what could normally be expected.

Benefits are those positive aspects of an alternative solution. They are the advantages, the strong points, and the good elements in an alternative. They should definitely not be limited to those points which can be quantified, but should also include qualitative aspects.

Chief decision-maker (CDM) is the one for whom the study of the system is ultimately being done. He is the person or committee who has authority to change the system in accordance with the results of a system study.

Complexity is a function of the number and type of elements and their degree of interaction.

Confidence is a subjective judgment by the analyst of the probable accuracy of a particular system utility score reflecting the true (actual) utility to the system over the planning horizon. The four levels of accuracy are very confident, confident, little confidence, and no confidence. The appropriate level of confidence for each entry in turn is determined by the validity of the facts, estimates, opinions, experiments, etc., the analyst was able to gather to validate the various weighted scores.

Constraints are restrictions and/or requirements placed on a system which must be met in order for the system to be viable.

Control refers to how much internal control a system has for ensuring the continual attainment of system objectives. This can range from a system having no internal feedback devices to systems which have very effective feedback loops.

Costs are those negative aspects of an alternative solution. They are the disadvantages, the weak points, or the bad elements in an alternative. They should definitely not be limited to dollar costs, but should include qualitative aspects as well as other quantitative points.

Criteria are more specific aspects of overall system objectives tailored to the system design choice.

Deliberators are those people who constitute the early majority of new system adopters. Their general motto is "Be not the last to lay the old aside, nor yet the first by which the new is tried." They tend to watch new developments in a positive frame of mind.

Desired performance specifies the relative desirability of various levels of systems performance from an overall perspective.

Discounted utility is an explicit means of considering the accuracy behind each of the ratings of each alternative design. The subjective probabilities for each rating are multiplied by the respective systems utility score to give the discounted utility score.

Discounted value is an overall indication of the utility or value to the system, considering the accuracy of the information gathered. It is given by adding the discounted utility scores for each alternative under the respective D column.

Economic criteria are measures of systems outcome in terms of costs which are usually expressed in monetary terms.

Egoistic needs are concerned with self-esteem and reputation in the way of self-confidence, independence, achievement, competence, knowledge, status, recognition, etc.

Environment includes all those factors which have an influence on the effectiveness of a system, but which are not controllable.

Environmental influence concerns how much affect the environmental conditions have on the functioning of the system. If the system is independent of the environment, it is called a closed system. If the system output is very much influenced by changes in the environment, the system is called open.

Exceptional performance is the level of performance indicated by the top 10%, or 90th to 100th percentile. This is well beyond what could normally be expected from systems designs in this category.

Evaluation report is used to establish the relative cost-effectiveness of each solution.

External constraints are restrictions placed on the system which must be met, but are not within the authority of the CDM to change.

Feasibility report is used to see if there are any workable solutions.

Feasible solution is a systems design which will solve the stated problem and will meet all constraints (restrictions).

Feedback is the comparing of actual systems performance with expected performance. Based on this comparison, a readjustment is made of system factors to come closer to the desired results.

Final report is the written summary identifying the problem, showing which solution is best, and establishing system goals for implementation.

Forward-lookers constitute the early adopters who are generally among the first 16 percent of the people who adopt an innovation. These people are generally not obsessed with trying new things, but are constantly looking and evaluating better ways to do things. They are generally well respected for their judgment on new ideas.

Fundamental causes of problems are the primary factors which are creating the difference between expected and actual system performance.

General problem-solving method is a generalized way for proceeding in the determination of the solution(s) to a problem, which is applicable to more than one class of problems.

Good performance are those outcomes which give better than the normal or usual performance in a particular situation.

Hierarchy concerns the relationship between systems and their components in terms of supra- and subordination.

Innovators are those people who first accept an idea. They have almost an obsession to try new things. These venturesome people are daring and possibly rash.

Interfaces are those boundaries where two systems meet, such that the output of one system is the input to the other. These boundaries can be internal and/or external to the system itself.

Internal constraints are restrictions placed on the system which must be met, but are within the authority of the CDM to change.

Objectives are the goals or results that the chief decision-maker wants, or should want, to attain in regard to a particular system.

Operational decisions are those decisions which involve the carrying out of the system objectives while keeping the system within constraint limits.

Optimizing is a method of problem-solving in which all viable (feasible) alternatives are compared. The best of these, in relation to the factors in the study, is accepted as the desired solution. Consequently, there is no better system solution.

Overall confidence is a means of showing what the estimated accuracy is overall in rating each alternative. It is determined by dividing the discounted value by the total value for each alternative.

Performance is the expected outcome of a particular alternative if it were implemented.

Performance measures are yardsticks which explicitly show in quantitative and qualitative terms what is barely acceptable, below average, average, above average, and exceptional systems performance.

Physiological needs relates to human requirements for living to include water, food, air, sleep, sex, shelter for protection from elements, disease, etc.

Planning horizon is the assumed time period for which any system solution must be effective in order to be acceptable.

Planning system change is preparation to learn how to implement the problem solution to ensure the desired effect.

Political criteria are a measurement of systems outcome in terms of power characteristics which are usually expressed in qualitative terms.

Potential solutions are methods which seem likely to solve the problem that has been identified with a particular system.

Preliminary report is prepared in order to determine what the problem is.

Problems are situations where there is a deviation between what is expected and what actually is.

Problem complexity is a function of the degree of abstraction of the system and its inherent complexity.

Problem diagnosis determines the cause or nature of the problem or situation.

Problem identification is the determination of what the primary question is that should be addressed in making a system more effective.

Problem-solving is development of a solution to attain what is expected or desired from what actually is.

Problem-solving method is a specific way for proceeding in the determination of the solution(s) to a certain type problem.

Psychological criteria are a measurement of systems outcome in terms of human characteristics usually expressed in qualitative terms.

Purposes of a system are determined by its relationship with the environment.

Relative rating is the resultant comparison by the analyst of his best estimate of what the systems performance would be if the particular alternative under consideration were implemented (i.e., systems simulation chart), when compared to the level of performance that could normally be expected (i.e., system utility function).

Safety needs are concerned with protection against danger, threats, deprivation, etc.

Satisficing is a method of problem-solving in which the first alternative that meets all restrictions (constraints) is accepted as the desired solution.

Scenario is a collection of estimates of the future status of major system contextual factors.

Scope of study specifies to what breadth and depth the system study will go.

Select alternative resolves which system solution is best by considering the overall performance of each alternative design and the explicit tradeoffs and risks involved with each choice.

Self-fulfillment needs are concerned with self-actualization through self-development, creativity, reaching one's potential, etc.

Simulations are used to estimate the expected performance of a system, using a model of the actual situation.

Skeptics are those people who constitute the late majority of adopters who approach innovations with a cautious air. They generally will not adopt an innovation until public opinion (or other companies, executives) definitely favor it.

Social needs relate to the desire for belonging, for association, for acceptance by one's fellows, friendship, love, etc.

Solutions are ways or methods for solving the system problem that has been identified.

Strategic decisions are those decisions which involve the determination of what the overall system objectives should be. It basically tries to answer the question of what the purpose of the system is, or where the system should be going?

Structure depicts the relationship between the components of the system to include its organization and interactions.

Study resources are the quantities of time, manpower, dollars, etc., which are allocated for the solving of a problem.

Symptoms are indications that the system isn't performing as expected or desired.

Systems are any set of components which could be looked at as working together for the overall objective of the whole.

System boundaries comprise that set of components which can be directly influenced or controlled in a system design.

System complexity is a function of the number and type of components and the level and degree of their interaction.

System context ascertains what the most relevant characteristics are of the system that encompasses the problem area.

System design is concerned with the appropriate selection of system components and their arrangement (structure), so as to meet the overall objectives of the system.

System goals are established once the solution is implemented and are what the expected level of system performance should be.

System objectives are desired standards of performance for the overall system.

System problems are deviations of system performance from what is expected compared to what actually is.

System study is a report on what the system problem is and what should be done about it.

System tradeoffs concerns the explicit recognition that a system can't be designed so as to satisfy the multiple, conflicting objectives and purposes of a system equally well.

System utility is a measure of the contribution of performance on a particular dimension to the total utility of the system, if a specific alternative were implemented. It is determined by multiplying the rating times the relative importance of the particular criterion.

Tactical decisions are those decisions which involve the determination of how to best accomplish the overall system objectives.

Technical criteria are measurements of systems outcome in terms of functional or operational characteristics usually expressed in quantitative terms.

Time to decision is the latest point in time in which the decision can be delayed without adversely affecting the system being studied.

Total value is an overall estimation of the relative utility or value to the system, if a particular alternative system design were implemented. It is determined by adding the scores under the U column or system utility score for each alternative.

Traditionalists are those people who are the last to accept new ideas. They are quite dubious of change, and tend to favor very strongly the status quo. Their attention seems fixed on the past rather than on the new and modern.

Wholeness is concerned with the overall aspects of a system, not with the individual components, per se.

Bibliography

Ackoff, Russell L., *A Concept of Corporate Planning.* New York: Wiley-Interscience, 1970.

———, "Science in the Systems Age: Beyond IE, OR, and MS," *Operations Research* (May-June 1973), pp. 661-671.

——— "Beyond Problem Solving," *General Systems,* (1974), XIX, 237-239.

———, *The Art of Problem Solving.* New York: John Wiley & Sons, 1978.

Alexander, C., *Notes on the Synthesis of Form.* Cambridge, MA: Harvard University Press, 1966.

Alinsky, Saul, *Rules for Radicals.* New York: Random House (Vintage Books), 1971.

Allen, Joseph, and Bennet Lientz, *Systems in Action.* Santa Monica, CA: Goodyear Publishing, 1978.

Ansoff, H., *Corporate Strategy.* New York: McGraw-Hill, 1965.

Anthony, Robert, *Planning and Control Systems.* Cambridge, MA: Harvard University Press, 1965.

Ashby, W. Ross, *Design for a Brain.* London: Chapman and Hall, 1952.

Athey, Thomas H., "Basic Methods of Information Gathering," *Journal of Systems Management* (January 1980).

———, "Systems People Need Much Better Problem Solving Methodologies," Keynote Speech, *Western Systems Conference,* October 1978.

———, "Fundamental Systems Concepts," *Journal of Systems Management* (November 1977).

———, "Training the Systems Analyst to Solve Complex Real-World Problems," *Proceedings of the 14th Annual Computer Personnel Research Conference.* Washington, D.C.: Association for Computer Machinery, 1976.

———, "The Development and Testing of a Seminar for Increasing the Cognitive Complexity of Individuals," Doctoral dissertation, Graduate School of Business Administration, University of Southern California, 1976.

———, "A Systems Methodology Useful to Business and Government," *Society for General System Research: Proceedings of the 6th Annual Far West Regional Conference.* San Jose, CA: Society for General System Research, 1975.

Awad, Elias, *Systems Analysis and Design.* Homewood, IL: Irwin, 1979.

Barnard, Chester I., *The Functions of the Executive.* Cambridge, MA: Harvard University Press, 1938.

Beer, Stafford, *Brain of the Firm: A Development in Management Cybernetics.* Frankfurt, Germany: Herder and Herder, 1972.

Biggs, Charles, et al. *Managing the Systems Development Process.* Englewood Cliffs, NJ: Prentice-Hall, 1980.

Blake, R., and J. Mouton, *Managerial Grid.* Houston, TX: Gulf, 1964.

Booth, Grayce, *The Distributed System Environment.* New York: McGraw-Hill, 1981.

Burke, James, *Connections.* London: MacMillan, 1978.

Carnegie, Dale, *How to Win Friends and Influence People.* New York: Simon & Schuster, 1969.

Chandler, Alfred, *Strategy and Structure.* Cambridge, MA: MIT Press, 1962.

Churchman, C. West, *The Systems Approach.* New York: Dell, 1968.

———, *Systems Approach and Its Enemies.* New York: Basic Books, 1979.

Cleland, D., and W. King, *Systems, Organizations, Management.* New York: McGraw-Hill, 1969.

Cyert, R., and J. March, *A Behavioral Theory of the Firm.* Englewood Cliffs, NJ: Prentice-Hall, 1964.

Dahl, Robert, *After the Revolution.* New Haven: Yale, 1962.

Dantzig, George, *Linear Programming and Extensions.* Corporation Research Studies, Princeton, NJ: Princeton University Press, 1963.

DeMarco, Tom, *Structured Analysis and System Specification.* New York: Yourdon Press, 1978.

Drucker, Peter, *Management: Tasks, Responsibilities, Practices.* New York: Harper and Row, 1974.

——, *The Effective Executive.* New York: Harper & Row, 1966.

——, *The Concept of the Corporation.* New York: New American Library (Mentor), 1964.

——, *Managing in Turbulent Times.* New York: Harper & Row, 1980.

Durant, Will and Ariel, *Lessons of History.* New York: Simon & Schuster, 1968.

Evan, Christopher, *The Micro Millenium.* New York: The Viking Press, 1980.

Forrester, J., *Principles of Systems.* Cambridge, MA: MIT Press, 1968.

Galbraith, J.K., *Economics and the Public Purpose.* Boston: Houghton Mifflin, 1973.

Gane, C. and T. Sarson, *Structured Systems Analysis.* Englewood Cliffs, NJ: Prentice-Hall, 1979.

Geradin, Lucien, *Bionics.* London: Weidenfeld & Nicolson (World University Library), 1968.

Halberstam, David, *The Best and the Brightest.* New York: Fawcett Crest Book, 1973.

Hanan, Mack, et al., *Sales Negotiation Strategies.* New York: Amacom, 1977.

Hardin, Garrett, *Nature and Man's Fate.* New York: New American Library (Mentor), 1959.

Haughen, J., and P. Binzen, *The Wreck of the Penn Central.* Boston: Little, Brown, 1971.

Horngren, Charles, *Accounting for Management Control.* Englewood Cliffs, NJ: Prentice-Hall, 1970.

Huff, Darrell, *How to Lie with Statistics.* New York: Norton, 1954.

Institute for Advanced System Studies, *Models for Training Environmental Education Teachers* (revised 1st report). 1976.

Jay, Anthony, *Management and Machiavelli.* New York: Holt, Rinehart & Winston, 1967.

——, *Corporation Man.* New York: Random House, 1971.

Jensen, Randall, and Charles Tonies, *Software Engineering.* Englewood Cliffs, NJ: Prentice-Hall, 1979.

Kast, F., and J. Rosenzweig, *Organization and Management: A Systems Approach.* New York: McGraw-Hill, 1974.

Katz, D., and R. Kahn, *The Social Psychology of Organizations.* New York: Wiley, 1966.

Keen, Peter G.W., "Information Systems and Organizational Change," *Communications of the ACM* (January 1981), pp. 24-33.

Keen, Peter G.W., and Michael Morton, *Decision Support Systems.* Reading, MA: Addison-Wesley, 1978.

Kotler, Philip, *Marketing and Management.* Englewood Cliffs, NJ: Prentice-Hall, 1972.

Lecht, Charles, *The Waves of Change.* New York: McGraw-Hill, 1977.

Letterman, Elmer, *Personal Power Through Creative Selling.* New York: Macmillan (Collier Books), 1967.

Levitt, Theodore, *The Marketing Mode.* New York: McGraw-Hill, 1969.

Levy, H., and M. Sarnat, *Investment and Portfolio Analysis.* New York: Wiley, 1972.

Lindblom, D. E., "The Science of Muddling Through," *Public Administration Review,* 19 (1959), 79-88.

Lucas, Henry Jr., *The Analysis, Design, and Implementation of Information Systems.* New York: McGraw-Hill, 1981.

Magruder, Jeb Stuart, *An American Life: One Man's Road to Watergate.* New York: Simon & Schuster (Pocket Books), 1975.

Maltz, Maxwell, *Psycho-Cybernetics.* New York: Simon & Schuster (Pocket Books), 1970.

Maslow, Abraham, *Motivation and Personality.* New York: Harper & Row, 1954.

McLean, Ephraim, and John Soden, *Strategic Planning for MIS.* New York: John Wiley & Sons, 1977.

McLuhan, Marshall, *Understanding Media: The Extensions of Man.* New York: New American Library (Mentor), 1973.

Meadow, Dennis, et al., *The Limits to Growth.* New York: Universe Books, 1972.

Molloy, John, *Dress for Success.* New York: Warner Books, 1976.

Morris, William, "On the Art of Modeling," *Management Science* (August 1967). pp. 13707-717.

Mort, Terry, *How to Influence the Buying Decision Process.* New York: Amacom, 1977.

Murdick, Robert G., *MIS Concepts and Design.* Englewood Cliffs, NJ: Prentice-Hall, 1980.

Murphy, John (editor), *Secrets of Selling.* New York: Dell, 1969.

Murphy, Walter, *Elements of Judicial Strategy.* Chicago: University of Chicago Press, 1964.

Nierenberg, Gerard, *Fundamentals of Negotiating.* New York: Elsevie-Dutton (Hawthorn Books), 1973.

Nierenberg, G., and H. Calero, *How to Read a Person Like a Book.* New York: Cornerstone Library, 1971.

Ogilvy, David, *Confessions of an Advertising Executive.* New York: Dell, 1963.

Osborn A., *Applied Imagination* (3rd ed.), New York: Scribner, 1979.

Ouchi, William, *Theory Z.* Reading, MA: Addison-Wesley, 1981.

Parsegian, V., *This Cybernetic World.* New York: Doubleday (Anchor Books), 1973.

Prell, Donald, "$C^{12}(DF)^2$," speech given at the University of Southern California, Los Angeles, October 1971.

Quade, E. S., *Systems Analysis Techniques for PPBS.* New York: Rand Corp., 1966.

Restak, Richard, *The Brain: The Last Frontier.* New York: Warner Books, 1979.

Richards, M., and P. Greenlaw, *Management: Decisions and Behavior.* Homewood, IL: Irwin, 1972.

Rogers, Everett, *Diffusion of Innovations.* New York: Macmillan, 1962.

Rush, Harold, *Behavioral Science: Concept and Management Application.* Report SPP216, New York: Conference Board, 1969.

Sackman, Harold, *Rudiments of a Real-World Theory of Man-Computer Problem Solving.* New York: Rand Corporation, 1974.

Sagan, Carl, *The Dragons of Eden.* New York: Random House, 1977.

Schein, Edgar, *Organizational Psychology.* Englewood Cliffs, NJ: Prentice-Hall, 1970.

Schneiderman, Ben, *Software Psychology: Human Factors in Computer and Information Systems.* Cambridge, MA: Winthrop Publishers, 1980.

Schoderbek, Peter, et al., *Management Systems: Conceptual Considerations.* (revised ed.) Dallas, TX: Business Publications, 1980.

Shaw, Gary, *Meat on the Hoof.* New York: Dell, 1972.

Shell, Richard L. and David F. Stelzer, *Business Horizons,* 14, no. 6 (December 1971), pp. 67-72.

Sides, Sharon S., "Determining My Best Life-Style," system study, California State Polytechnic University, Pomona, May 1976.

Siegel, G. B., *The Unity of the Systems Approach* (manuscript). Los Angeles: University of Southern California, 1969.

Slonim, Morris, *Sampling*. New York: Simon & Schuster, 1966.

Steiner, George, *Strategic Planning*. New York: The Free Press, 1979.

Strauss, Nicholas, "Selecting Best Car for a Fleet Purchase," system study, California State Polytechnic University, Pomona, March 1974.

Talbot, Gregory, "Selecting Best Parachute for Sky Diving," system study, California State Polytechnic University, Pomona, December 1975.

Tharrington, Roy B., "Best Method for Setting Newspaper Type," system study, California State Polytechnic University, Pomona, December 1972.

Thompson, James, *Organizations in Action*. New York: McGraw-Hill, 1967.

Toffler, Alvin, *Future Shock*. New York: Random House, 1970.

——, *The Futurists*. New York: Random House, 1972.

——, *The Third Wave*. New York: Bantam Books, 1981.

Towards Postal Excellence. Report of the President's Commission on Postal Organization, 1968, Government Printing Office.

Van Horne, James, *Financial Management and Policy*. Englewood Cliffs, NJ: Prentice-Hall, 1971.

Vickers, Geoffrey, *Value Systems and the Social Process*. New York: Basic Books, 1968.

Warfield, John, *Societal Systems: Planning, Policy, and Complexity*. New York: John Wiley & Sons, 1976.

Webb, Eugene, et al., *Unobtrusive Measures: Nonreactive Research in the Social Sciences*. Chicago: Rand McNally, 1966.

Weinberg, Gerald, *An Introduction to General Systems Thinking*. New York: John Wiley & Sons, 1975.

White, Ron, "Best Way to Raise a Child," system study, California State Polytechnic University, Pomona, May 1977.

Wildavsky, Aaron, *The Politics of the Budgetary Process*. Boston: Little, Brown, 1964.

Williams, J., *The Compleat Strategyst* (revised ed.). New York: McGraw-Hill, 1965.

Wilson, Edward, *Sociobiology* (abridged ed.). Cambridge, MA: Harvard University Press, 1980.

Yourdon, Edward, *Managing the Structured Techniques*. New York: Yourdon Press, 1979.

Zwicky, Fritz, *Morphology of Propulsive Power*. Pasadena, CA: Society for Morphological Research, 1962.

Index

*Terms in the glossary

T